Collins

AQA GCSE

Biology

Biology

AQA GCSE

Revision Guide

Ian Honeysett

Contents

	Revise	Practise	Review
Recap of KS3 Key Concepts			p. 6 ☐

Paper 1 Cell Biology

Cell Structure	p. 8 ☐	p. 26 ☐	p. 56 ☐

A Typical Animal Cell
Plant Cells
Prokaryotic and Eukaryotic Cells
A Typical Bacterial Cell

Investigating Cells	p. 10 ☐	p. 26 ☐	p. 57 ☐

The Size of Cells
Using Microscopes to Look at Cells
Calculating Magnification
Growing Microorganisms

Cell Division	p. 12 ☐	p. 28 ☐	p. 58 ☐

Chromosomes
Mitosis and the Cell Cycle
Stem Cells
Uses of Stem Cells

Transport In and Out of Cells	p. 14 ☐	p. 28 ☐	p. 59 ☐

Diffusion
Factors Affecting Diffusion
Osmosis
Active Transport
Comparing Processes

Paper 1 Organisation

Levels of Organisation	p. 16 ☐	p. 30 ☐	p. 60 ☐

Specialised Cells
Tissues, Organs and Systems

Digestion	p. 18 ☐	p. 30 ☐	p. 60 ☐

Enzymes
Enzymes in Digestion
Bile and Digestion

Blood and the Circulation	p. 20 ☐	p. 32 ☐	p. 61 ☐

Blood
Blood Vessels
The Heart
Gaseous Exchange

HT Higher Tier Content

Contents

	Revise	Practise	Review
Non-Communicable Diseases	p. 22	p. 32	p. 62
Health and Disease			
Risk Factors			
Diseases of the Heart			
Cancer			
Transport in Plants	p. 24	p. 33	p. 62
Plant Tissues			
Water Transport			
Translocation			

Paper 1   **Infection and Response**

	Revise	Practise	Review
Pathogens and Disease	p. 34	p. 64	p. 94
Pathogens and Disease			
Viral Pathogens			
Bacterial Diseases			
Protists and Disease			
Fungal Diseases			
Human Defences Against Disease	p. 36	p. 64	p. 94
Preventing Entry of Pathogens			
The Immune System			
Boosting Immunity			
Treating Diseases	p. 38	p. 66	p. 95
Antibiotics			
Developing New Drugs			
HT *Monoclonal Antibodies*			
Plant Disease	p. 40	p. 67	p. 97
HT *Detecting and Identifying Plant Disease*			
Examples of Plant Diseases			
Plant Defences			

Paper 1   **Bioenergetics**

	Revise	Practise	Review
Photosynthesis	p. 42	p. 68	p. 98
Photosynthesis			
Factors Affecting Photosynthesis			
Converting Glucose			
Respiration and Exercise	p. 44	p. 68	p. 99
The Importance of Respiration			
Aerobic Respiration			
Anaerobic Respiration			
Exercise and Respiration			
Metabolism			

HT Higher Tier Content

Contents

		Revise	Practise	Review
Paper 2	**Homeostasis and Response**			
Homeostasis and Body Temperature		p. 46 ☐	p. 70 ☐	p. 100 ☐
The Importance of Homeostasis				
Control Systems				
Control of Body Temperature				
The Nervous System and the Eye		p. 48 ☐	p. 70 ☐	p. 100 ☐
The Nervous System				
The Brain and the Eye				
Hormones and Homeostasis		p. 50 ☐	p. 71 ☐	p. 101 ☐
The Endocrine System				
Control of Blood Glucose				
Water Balance				
Hormones and Reproduction		p. 52 ☐	p. 72 ☐	p. 103 ☐
The Sex Hormones				
Control of the Menstrual Cycle				
Reducing Fertility				
HT *Increasing Fertility*				
Plant Hormones		p. 54 ☐	p. 73 ☐	p. 103 ☐
Functions of Plant Hormones				
HT *Uses of Plant Hormones*				
Paper 2	**Inheritance, Variation and Evolution**			
Sexual and Asexual Reproduction		p. 74 ☐	p. 104 ☐	p. 112 ☐
Asexual Reproduction				
Sexual Reproduction and Meiosis				
Asexual Versus Sexual Reproduction				
DNA and Protein Synthesis		p. 76 ☐	p. 105 ☐	p. 113 ☐
The Genome				
The Structure of DNA				
Making Proteins				
HT *Mutations*				
Patterns of Inheritance		p. 78 ☐	p. 105 ☐	p. 113 ☐
Gregor Mendel				
Modern Ideas About Genetics				
Genetic Crosses				
Genetic Disorders				
Sex Determination				

HT Higher Tier Content

Contents

	Revise	Practise	Review
Variation and Evolution	p. 80	p. 106	p. 114
Variation			
Natural Selection			
Evidence for Evolution			
Manipulating Genes	p. 82	p. 107	p. 115
Selective Breeding			
Genetic Engineering			
Cloning			
Classification	p. 84	p. 107	p. 115
Principles of Classification			
Extinction			
Evolutionary Trees			
Speciation			

Paper 2 **Ecology**

	Revise	Practise	Review
Ecosystems	p. 86	p. 108	p. 116
Relationships Between Organisms			
Adaptations			
Studying Ecosystems			
Cycles and Feeding Relationships	p. 88	p. 109	p. 117
Decomposition			
Recycling Materials			
Feeding Relationships			
Disrupting Ecosystems	p. 90	p. 110	p. 117
Biodiversity			
Pollution			
Overexploitation			
Conserving Biodiversity			
Feeding the World	p. 92	p. 111	p. 119
The Need for More Food			
Manipulating Energy Flow			
Biotechnology			
Mixed Exam-Style Questions	p. 120		
Answers	p. 132		
Glossary and Index	p. 140		

HT Higher Tier Content

Review Questions

Recap of KS3 Key Concepts

1 What is the basic unit that all living things are made of? [1]

2 The diagrams show an animal cell and a plant cell.

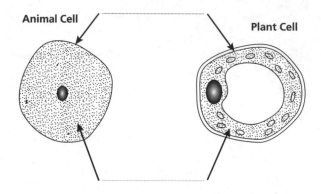

Animal Cell

Plant Cell

a) Some structures are found in both types of cell.

Label **two** of these structures, shown by the arrows, on the diagrams. [2]

b) Name **two** structures that are present in plant cells but **not** in animal cells. [2]

3 Human ova (eggs) and human sperm have important roles in reproduction.

a) What are eggs and sperm? [1]
Tick **one** box.

tissues ☐   cells ☐   organs ☐

b) What does a sperm use to swim towards an egg? [1]

c) Name the male reproductive organ where sperm are made. [1]

d) In reproduction, a sperm fuses with an egg.

What is this process called? [1]

4 Carbon monoxide, nicotine and tar all get into the lungs when a person smokes.

Write down **one** harmful effect on the body of **each** of these substances. [3]

5 What is the process called by which plants produce glucose and oxygen? [1]

6 What is the process called by which the human body releases energy from glucose and oxygen? [1]

7 Give **two** functions of the human skeleton. [2]

8 The chart shows a way to group living organisms.

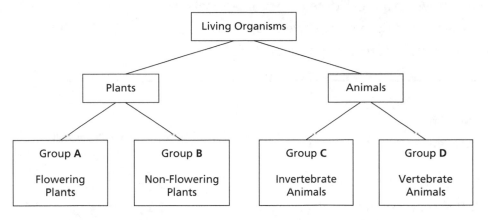

a) To which group, **A**, **B**, **C** or **D**, do the following living organisms belong?

    i) Flatworm   **ii)** Mouse   **iii)** Daffodil   **iv)** Human   **[4]**

b) Some flatworms are parasites.

    What is a parasite?   **[1]**

9 Rachel cut her knee playing netball. A medic put a plaster over the cut.

a) A plaster helps to stop a cut getting infected.

    Give the name of **one** type of microorganism that can infect a cut.   **[1]**

b) While she was cleaning Rachel's knee, the medic wore rubber gloves.

    Why is wearing rubber gloves important for the medic's health?   **[1]**

10 The human lungs contain millions of alveoli.

a) i) Which gas passes into the blood from the air in the lungs?   **[1]**

    ii) Which gas passes out of the blood into the air in the lungs?   **[1]**

b) The walls of the capillaries and the alveoli are very thin.

    Why do they need to be thin?   **[1]**

c) There are millions of alveoli in the lungs. They provide a very large surface area.

    Why is a large surface area important?   **[1]**

Total Marks _____ / 27

Cell Structure

You must be able to:

- Describe the structure of a typical animal cell
- Describe how a plant cell differs from an animal cell
- Recall the main differences between prokaryotic and eukaryotic cells
- Describe the structure of a typical bacterial cell.

A Typical Animal Cell

- All cells have structures inside them – these are called **sub-cellular structures**.
- In an animal cell, the sub-cellular structures include:

  - a **nucleus**, which controls the activities of the cell and contains the genetic material
  - **cytoplasm**, in which most of the chemical reactions take place
  - a **cell membrane**, which controls the passage of substances into and out of the cell
  - **mitochondria**, where aerobic respiration takes place
  - **ribosomes**, where proteins are synthesised (made).

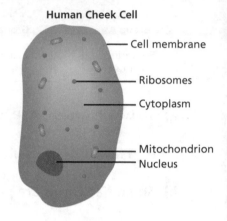

Human Cheek Cell

- Cell membrane
- Ribosomes
- Cytoplasm
- Mitochondrion
- Nucleus

Plant Cells

- Plant cells and algal cells contain all the sub-cellular structures found in animal cells.
- They also have:
  - a **cell wall** made of **cellulose**, which strengthens the cell
  - a permanent **vacuole** filled with cell sap, which supports the plant.
- Plants need to make their own food, so some of their cells contain chloroplasts.
- **Chloroplasts** absorb light to make food (glucose) by photosynthesis (see pages 42–43).

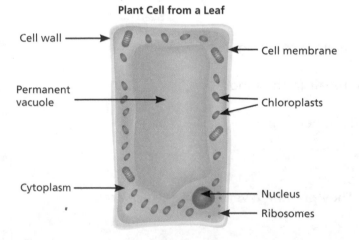

Plant Cell from a Leaf

- Cell wall
- Cell membrane
- Permanent vacuole
- Chloroplasts
- Cytoplasm
- Nucleus
- Ribosomes

(see pages 42–43)

Key Point

Not all plant cells have chloroplasts. For example, they are not present in root cells because they do not receive any light.

Prokaryotic and Eukaryotic Cells

- There are two main types of cell:
  - prokaryotic
  - eukaryotic.
- Plant, animal and fungal cells are all eukaryotic.
- Bacterial cells are prokaryotic.
- There are a number of differences between the two types of cell.
- Prokaryotic cells are much smaller in size and:
  - the genetic material is not enclosed in a nucleus
  - the genetic material is a single DNA loop and there may be one or more small rings of DNA, called **plasmids**
  - they do not contain mitochondria or chloroplasts.

Key Point

Prokaryotic cells are much simpler in structure than eukaryotic cells. That is why scientists think that they developed before eukaryotic cells.

A Typical Bacterial Cell

- Bacterial cells have many different shapes – some are round, some are rod-shaped and some are spiral – but they are all prokaryotic cells.
- In bacterial cells, the roles of mitochondria and chloroplasts are taken over by the cytoplasm.
- Plasmids are present, which are circles of DNA that can be transferred from one cell to another.
- Plasmids allow bacterial cells to move genes from one cell to another.

Key Point

Plasmids have become very useful to scientists. They allow genes to be inserted into bacteria in genetic engineering (see pages 82–83).

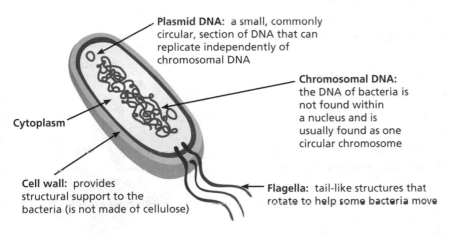

A Typical Bacterial Cell

Plasmid DNA: a small, commonly circular, section of DNA that can replicate independently of chromosomal DNA

Chromosomal DNA: the DNA of bacteria is not found within a nucleus and is usually found as one circular chromosome

Cytoplasm

Cell wall: provides structural support to the bacteria (is not made of cellulose)

Flagella: tail-like structures that rotate to help some bacteria move

Key Words

sub-cellular structures
nucleus
cytoplasm
cell membrane
mitochondria
ribosomes
cell wall
cellulose
vacuole
chloroplast
prokaryotic
eukaryotic
plasmid

Quick Test

1. Which sub-cellular structure controls the activities inside the cell?
2. Where are proteins made in a cell?
3. Write down **three** structures that are found in plant cells but **not** in animal cells.
4. What is the function of cell sap?
5. Where is DNA found in a bacterium?

Investigating Cells

You must be able to:

- Use different units of measurement to describe the size of cells
- Describe the advantages of using an electron microscope to view cells
- Analyse images of cells and perform calculations involving magnification, actual size and image size
- Describe the techniques used for culturing microorganisms.

The Size of Cells

- A typical plant cell may be about 0.1mm in diameter and an animal cell 0.02mm in diameter.
- Prokaryotic cells are smaller – often about 0.002mm long.
- To describe the size of cells and sub-cellular structures, scientists use units that have different prefixes.

Unit	Number of Units in One Metre (1m)	
centimetre (cm)	100	1×10^2
millimetre (mm)	1 000	1×10^3
micrometre (μm)	1 000 000	1×10^6
nanometre (nm)	1 000 000 000	1×10^9

Using Microscopes to Look at Cells

- It is not possible to see cells as separate objects using the naked eye.
- The ability to see two or more objects as separate objects is called **resolution**.
- The light microscope was developed in the late 16th century and gave a greater resolution than the human eye.
- It allowed scientists to see plant, animal and bacterial cells.
- Some sub-cellular structures are even smaller than the resolution achieved by a light microscope and cannot be seen using this method.
- In 1933, scientists first used an **electron microscope**.
- An electron microscope passes electrons, rather than light, through the specimen and can give much better resolution.
- Cells can be seen in much finer detail, e.g.
  - the structures inside mitochondria and chloroplasts can be studied and this has helped scientists to find out how they work
  - ribosomes can be seen and their role in making proteins can be studied.

> **Key Point**
>
> The size of cells is often given in micrometres, e.g. a plant cell may be 100μm, an animal cell 20μm and a bacterium 2μm in size.

A Light Microscope

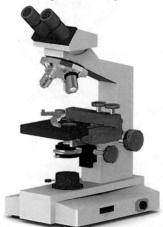

An Electron Microscope

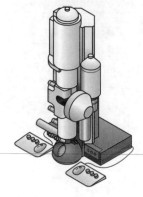

REQUIRED PRACTICAL	
Use a light microscope to observe, draw and label a selection of plant and animal cells.	
Sample Method 1. Place a tissue sample on a microscope slide. 2. Add a few drops of a suitable stain. 3. Lower a coverslip onto the tissue. 4. Place the slide on the microscope stage and focus on the cells using low power. 5. Change to high power and refocus. 6. Draw any types of cells that can be seen. 7. Add a scale line to the diagram.	**Considerations, Mistakes and Errors** • The scale line can be added by focusing on the millimetre divisions of a ruler.

Calculating Magnification

- When a microscope is used to look at cells, scientists will often take photographs or produce drawings.
- These images are many times larger than the real cell or structure.
- The **magnification** is how many times larger the image is than the real object.

$$\text{magnification} = \frac{\text{size of image}}{\text{size of real object}}$$

Growing Microorganisms

- Bacteria use a type of simple cell division, called **binary fission**, to multiply. This is an example of asexual reproduction (see pages 74–75).
- They can multiply as frequently as once every 20 minutes if they have enough nutrients and a suitable temperature.
- Bacteria can be grown in a nutrient broth solution or as colonies on a type of jelly called **agar**. This is called a **culture**.
- The agar is usually in the bottom of a flat dish called a **Petri dish**.
- Cultures of microorganisms need to be uncontaminated. This is important so that specific strains can be used to test the effects of antibiotics or disinfectants. Uncontaminated cultures can be produced by the following sterile technique:

1 The Petri dishes and agar must be sterilised before use to kill unwanted microorganisms.

2 An inoculating loop is sterilised by passing it through a flame.

3 The cooled inoculating loop is used to transfer bacteria to the agar.

Sterilising an Inoculating Loop

Hold inoculating loop in Bunsen flame until red hot

Bunsen burner

Petri dish

Agar

4 The lid of the Petri dish must be quickly removed and replaced when transferring bacteria and secured with tape to stop it coming off (otherwise microorganisms from the air may contaminate the culture).

5 The dish is stored upside down to stop condensation dripping onto the agar surface.

- This procedure is called aseptic technique and ensures that only the required microorganisms are grown.
- In school and college laboratories, cultures should be incubated at a maximum temperature of 25°C. This reduces the likelihood of the growth of harmful bacteria.
- In industry, higher temperatures can be used for more rapid growth.

Quick Test

1. Arrange these structures in order of size with the largest first:
   bacterium   liver cell   nucleus   ribosome
2. A nucleus is measured as 0.005mm in diameter. How many micrometres is this?
3. A student draws a cheek cell. The cell in their drawing is 50mm wide. In real life the cell is 0.025mm. What is the magnification of their drawing?
4. Write down **two** precautions that should be taken to stop a bacterial culture getting contaminated.

Key Point

Magnification is not the same as resolution. Light microscopes can be built to magnify more and more, but after a certain point the images do not get any clearer.

Key Point

Fungi can also be grown on agar in Petri dishes. Viruses cannot be grown on agar because they need living cells to reproduce. They have to be grown inside suitable host cells.

Key Point

In optimum conditions, bacteria can divide approximately every 20 minutes. This is the mean (average) division time.

The area of a colony of bacteria can be calculated using the equation:
area of a circle = πr^2.

Key Words

resolution
electron microscope
magnification
binary fission
agar
culture
Petri dish

Cell Division

You must be able to:

- Describe the arrangement of chromosomes in a cell
- Explain the importance of mitosis in the cell cycle
- Describe the properties and functions of stem cells
- Explain the use of stem cells.

Chromosomes

- The nucleus of a cell contains **chromosomes** made of **DNA**.
- Each chromosome carries hundreds to thousands of **genes**.
- Different genes contain the code to make different proteins and so control the development of different characteristics.
- In body cells, the chromosomes are found in pairs, with one chromosome coming from each parent.
- Different species have different numbers of pairs of chromosomes, e.g. humans have 23 pairs and dogs have 39 pairs.

A Section of One Chromosome

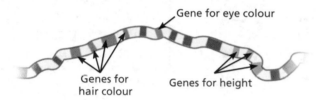

Gene for eye colour

Genes for hair colour

Genes for height

Mitosis and the Cell Cycle

- Cells go through a series of changes involving growth and division, called the **cell cycle**.
- One of the stages is **mitosis**, when the cell divides into two identical cells.
- Before a cell can divide, it needs to grow and increase the number of sub-cellular structures, such as ribosomes and mitochondria.
- The DNA then replicates to form two copies of each chromosome. In this way, the genetic material is doubled.
- During mitosis:
  1. One set of chromosomes is pulled to each end of the cell.
  2. The nucleus divides.
  3. The cytoplasm and cell membranes divide to form two identical cells.

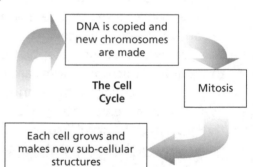

DNA is copied and new chromosomes are made

Mitosis

The Cell Cycle

Each cell grows and makes new sub-cellular structures

Mitosis

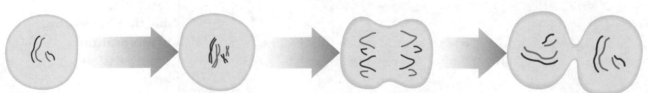

Parent cell with two pairs of chromosomes.

Each chromosome replicates (copies) itself.

Chromosomes line up along the centre of the cell, divide and the copies move to opposite poles.

Each 'daughter' cell has the same number of chromosomes, and contains the same genes, as the parent cell.

- Cell division by mitosis is important because it makes new cells for:
  - growth and development of multicellular organisms
  - repairing damaged tissues
  - asexual reproduction.

Stem Cells

- Some cells are **undifferentiated** – they have not yet become specialised.
- This means that they can divide to make different types of cells. They are called **stem cells**.
- Stem cells are found in human embryos, in the umbilical cord of a new born baby, and in some organs and tissues.
- Stem cells from human embryos are called **embryonic stem cells** and can make all types of cells.
- **Adult stem cells** are found in some organs and tissues, e.g. bone marrow. They can only make certain types of cells and their capacity to divide is limited.

Uses of Stem Cells

- Stem cells may be very useful in treating conditions where cells are damaged or not working properly, such as in diabetes and paralysis.
- They could be used to replace the damaged cells.
- A cloned embryo of the patient may be made and used as a source of stem cells. This is called **therapeutic cloning**.
- Stem cells from the cloned embryo will not be rejected by the patient's body, so they could be very useful in treating the patient.
- Some people are concerned about using stem cells from cloned embryos:
  - there may be risks, such as the transfer of viral infection
  - they may have ethical or religious objections.
- In plants, stem cells are found in special areas called **meristems**.
- These meristems allow plants to make new cells for growth.
- The stem cells can be used to produce clones of plants quickly.
- This could be useful for a number of reasons:
  - rare species can be cloned to protect them from extinction
  - large numbers of identical crop plants with special features, such as disease resistance, can be made.

> **Key Point**
>
> Although there are adult stem cells all over the body, they are very difficult to find and isolate.

Potential Uses for Stem Cells

Stem cells

Pancreatic islet cells

Bone marrow

Heart muscle

Blood

Neurones

> **Key Point**
>
> Stem cell use (and the different views for and against it) is a good example of an ethical argument concerning a new technology.

> **Key Words**
>
> **chromosomes**
> **DNA**
> **gene**
> **cell cycle**
> **mitosis**
> **undifferentiated**
> **stem cell**
> **embryonic stem cells**
> **adult stem cells**
> **therapeutic cloning**
> **meristems**

> **Quick Test**
>
> 1. How many chromosomes are there in each human body cell?
> 2. Why is it important that the chromosomes are copied before mitosis occurs?
> 3. What is a stem cell?
> 4. Why does a plant have a meristem at the tip of the shoot and the tip of each root?

Transport In and Out of Cells

You must be able to:

- Describe the process of diffusion
- Explain the factors involved in moving molecules in and out of cells
- Describe how water can move by osmosis
- Explain why some substances are moved by active transport
- Compare diffusion, osmosis and active transport.

Diffusion

- Many substances move into and out of cells, across the cell membranes, by **diffusion**.
- Diffusion is the net movement of particles from an area of higher concentration to an area of lower concentration until they are evenly spread out.
- This happens because the particles move randomly and spread out.
- There are many examples of diffusion in living organisms:
  - oxygen and carbon dioxide diffuse during gas exchange in lungs, gills and plant leaves
  - urea diffuses from cells into the blood plasma for excretion by the kidney
  - digested food molecules from the small intestine diffuse into the blood.

Factors Affecting Diffusion

- The factors that affect the rate of diffusion are:
  - the difference in concentrations, known as the **concentration gradient**
  - the temperature
  - the surface area of the membrane.
- A single-celled organism has a large **surface area to volume ratio**.
- This allows enough molecules to diffuse into and out of the cell to meet the needs of the organism.
- In multicellular organisms, there is a smaller surface area to volume ratio. However, surfaces and organ systems are specialised for exchanging materials.
- The small intestine and lungs in mammals, gills in fish, and the roots and leaves in plants, are all adapted for exchanging materials:
  - they have a large surface area
  - the surface is thin so that molecules only have to diffuse a short distance
  - surfaces are usually kept moist so that substances can dissolve and diffuse across cell membranes faster
  - in animals, a rich blood supply maintains the concentration gradient
  - in animals, ventilation occurs to speed up gaseous exchange.

Osmosis

- Water may move across cell membranes by **osmosis**.
- Osmosis is the diffusion of water from a dilute solution to a concentrated solution through a partially permeable membrane.

Diffusion

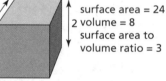

Some particles

High concentration — Many particles — Low concentration

Net movement of particles

The Effect of the Size of an Organism on its Surface Area to Volume Ratio

Organism A

surface area = 6
volume = 1
surface area to volume ratio = 6

Organism B

surface area = 24
volume = 8
surface area to volume ratio = 3

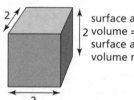

Key Point

A dilute solution contains lots of water. A concentrated solution contains less water.

Osmosis

| Dilute solution (high concentration of water) | Concentrated solution (low concentration of water) |

Partially permeable membrane

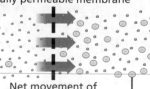

Net movement of water molecules

These molecules are too large to pass through the membrane

REQUIRED PRACTICAL	
Investigating the effect of different concentrations of sugar solution on plant tissue.	
Sample Method Potatoes can be used to measure the effect of sugar solutions on plant tissue: 1. Cut some cylinders of potato tissue and measure their mass. 2. Place the cylinders in different concentrations of sugar solution. 3. After about 30 minutes remove the cylinders and measure their mass again. If the cylinders change in mass, they have gained or lost water by osmosis.	**Considerations, Mistakes and Errors** • The cylinders need to be left in the solution long enough for a significant change in mass to occur. • Before the mass of the cylinders is measured again, they should be rolled on tissue paper to remove any excess solution.
Variables • The independent variable is the one deliberately changed – in this case, the concentration of sugar solution. • The dependent variable is the one that is measured – in this case, the change in mass of the potato. • The control variables are kept the same – in this case, the temperature, the length of time the cylinders were left in the solution and the volume of the solution.	**Hazards and Risks** • Care must be taken when cutting the cylinders of potato.

Key Point

If the potato cylinders do not lose or gain water, then the sugar solution must be the same concentration as the potato tissue.

Active Transport

- **Active transport** moves substances against a concentration gradient, from an area of low concentration to high concentration.
- This requires energy from respiration.
- Active transport allows mineral ions to be absorbed into plant root hairs from very dilute solutions in the soil.
- Active transport also allows sugar molecules to be absorbed from lower concentrations in the gut into the blood, which has a higher concentration.

A Cell Absorbing Ions by Active Transport

Root hair cell with high concentration of nitrate ions

Soil with low concentration of nitrate ions

Cell uses energy to 'pull' ions against the concentration gradient

Comparing Processes

	Diffusion	Osmosis	Active Transport
Allows molecules to move	✓	✓	✓
Movement is down a concentration gradient	✓	✓	✗
Always involves the movement of water	✗	✓	✗
Needs energy from respiration	✗	✗	✓

Key Point

Anything that stops respiration occurring, such as lack of oxygen or metabolic poisons, will stop active transport.

Quick Test

1. A person opens a bottle of perfume. Why do people in the room smell it faster on a warm day?
2. What is required for substances to be absorbed against a concentration gradient?
3. In addition to a large surface area, name **one** other feature that makes an exchange surface more efficient.

Key Words

diffusion
concentration gradient
surface area to volume ratio
osmosis
active transport

Levels of Organisation

You must be able to:

- Explain how cells become specialised for particular roles
- Describe examples of specialised cells
- Explain how cells can form tissues, organs and systems.

Specialised Cells

- Cells are the basic building blocks of all living organisms.
- As an organism develops, cells differentiate to form different types of cells. They become **specialised**.
- Most types of animal cell differentiate at an early stage, but many types of plant cell can differentiate throughout their life.
- As a cell differentiates:
  - it may change shape
  - different sub-cellular structures develop to let it to carry out a specific function.
- Specialised animal cells include sperm, nerve and muscle cells.

> ### Key Point
>
> If cells are specialised, they become more efficient at their job but may lose the ability to do other jobs.

A Sperm Cell

Tail: to propel the sperm to fertilise the egg

Mitochondria: sperm have many of these cell components, which are the major site of respiration, to provide energy for their journey

Nucleus: contains only one set of the genetic material

Acrosome: contains enzymes to allow the sperm to penetrate the outer layer of the egg

A Motor Neurone (Type of Nerve Cell)

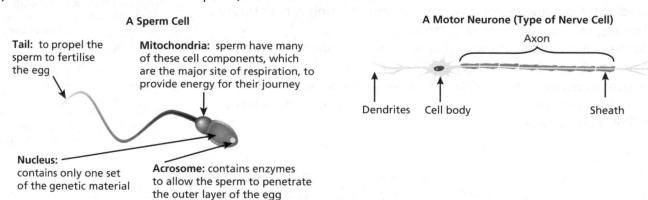

Axon

Dendrites   Cell body   Sheath

A Muscle Cell

Nucleus

Mitochondria

Protein fibres that can contract

Many mitochondria for energy

- In plants, root hair, xylem and phloem cells are all specialised cells.

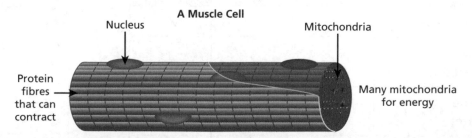

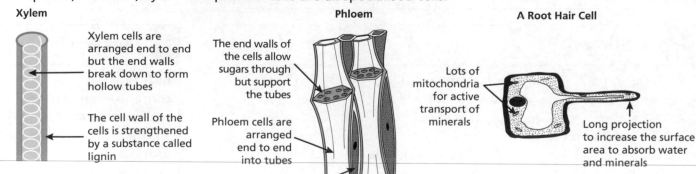

Xylem

Xylem cells are arranged end to end but the end walls break down to form hollow tubes

The cell wall of the cells is strengthened by a substance called lignin

Phloem

The end walls of the cells allow sugars through but support the tubes

Phloem cells are arranged end to end into tubes

Companion cell

A Root Hair Cell

Lots of mitochondria for active transport of minerals

Long projection to increase the surface area to absorb water and minerals

Tissues, Organs and Systems

- In most organisms, cells are arranged into **tissues**.
- A tissue is a group of cells with a similar structure and function, which all work together to do a job, e.g.
  - muscle tissue contracts to produce movement
  - glandular tissue produces substances such as enzymes and hormones
  - epithelial tissue covers organs.
- **Organs** are groups of different tissues, which all work together to perform a specific job.
- Each organ may contain several tissues.
- For example, the stomach is an organ that contains:
  - muscle tissue that contracts to churn the contents
  - glandular tissue to produce digestive juices
  - epithelial tissue to cover the outside and inside of the stomach.
- Organs are organised into **organ systems**, which are groups of organs working together to do a particular job.
- The digestive system is an example of an organ system, in which several organs work together to digest and absorb food.
- Lots of organ systems work together to make an organism.

Muscle tissue
Can contract to bring about movement

Glandular tissue
Can produce substances such as enzymes and hormones

Epithelial tissue
Covers all parts of the body

The Stomach

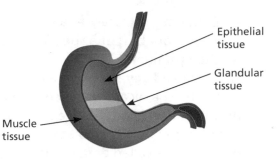

Epithelial tissue

Glandular tissue

Muscle tissue

The Digestive System

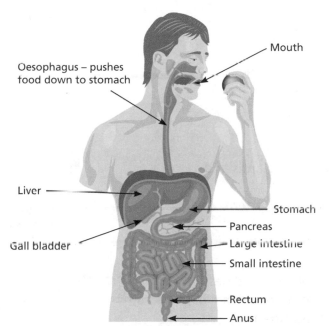

Mouth

Oesophagus – pushes food down to stomach

Liver

Stomach

Pancreas

Large intestine

Gall bladder

Small intestine

Rectum

Anus

Key Point

Single-celled organisms do not have tissues.

Some simple animals do not have organs, only tissues.

Quick Test

1. Why do cells differentiate?
2. Why do sperm cells contain lots of mitochondria?
3. What are a group of cells with a similar structure and function called?
4. What is the function of epithelia?
5. Is the heart a tissue, organ or organ system?

Key Words

specialised
tissue
organ
organ system

Digestion

You must be able to:

- Explain how enzymes work
- Describe the role of enzymes in digestion
- Explain how bile can speed up the digestion of lipids.

Enzymes

- Enzymes are biological **catalysts** – they speed up chemical reactions in living organisms.
- Enzymes have a number of properties:
  - They are all large proteins.
  - There is a space within the protein molecule called the **active site**.
  - Each enzyme catalyses a specific reaction.
  - They work best at a specific temperature and pH called the **optimum**.
- The '**lock and key theory**' is a model used to explain how enzymes work: the chemical that reacts is called the substrate (key) and it fits into the enzyme's active site (lock).
- High temperature and extremes of pH make enzymes change shape. This is called **denaturing**.
- The enzyme cannot work once it has been denatured, because the substrate cannot fit into the active site – the lock and key no longer fit together.

Enzymes in Digestion

- Digestive enzymes are produced by specialised cells in glands and in the lining of the gut:
  1. The enzymes pass out of the cells into the digestive system.
  2. They come into contact with food molecules.
  3. They catalyse the breakdown of large insoluble food molecules into smaller soluble molecules.
- The digestive enzymes, **protease**, **lipase** and **carbohydrase**, digest proteins, lipids (fats and oils) and carbohydrates to produce smaller molecules that can be easily absorbed into the bloodstream.

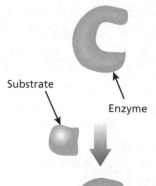

Substrate

Enzyme

Combined substrate and enzyme. Reaction can take place

Products

Substrate is broken down and enzyme can be reused

Key Point

The 'Lock and Key Theory' is an example of how models are used in science to try and explain observations.

Enzyme

Heat

Enzyme denatured by heat

REQUIRED PRACTICAL	
Use qualitative reagents to test for a range of carbohydrates, lipids and proteins.	
Sample Method 1. To test for sugars, e.g. glucose, add Benedict's reagent and heat in a water bath for two minutes. If sugar is present, it will turn red. 2. To test for starch add iodine solution. If starch is present, it will turn blue-black. 3. To test for protein add biuret reagent. If protein is present, it will turn purple.	**Considerations, Mistakes and Errors** • Do not boil the mixture for a long time, because any starch present might break down into sugar and test positive. • Refer to 'iodine solution' not 'iodine'. • Sometimes the purple colour is difficult to see. Try holding the test tube in front of a sheet of white paper.

- **Amylase:**
  - is produced in the salivary glands and the pancreas
  - is a carbohydrase that breaks down starch into sugar (maltose).
- Starch → maltose
- **Protease:**
  - is produced in the stomach, pancreas and small intestine
  - breaks down proteins into amino acids.
- Protein → peptides or amino acids
- **Lipase:**
  - is produced in the pancreas and small intestine
  - breaks down lipids (fats) into fatty acids and glycerol.
- Fats → fatty acids + glycerol
- These products of digestion can be used to build new carbohydrates, lipids and proteins.

Bile and Digestion

- **Bile** is a liquid made in the liver and stored in the gall bladder.
- It is alkaline to neutralise hydrochloric acid from the stomach.
- It also emulsifies fat to form small droplets, increasing the surface area for enzymes to act on.
- The alkaline conditions and large surface area increase the rate at which fat is broken down by lipase.

Liver, Stomach, Pancreas, Small intestine, Bile duct, Gall bladder

REQUIRED PRACTICAL	
Investigate the effect of pH on the rate of reaction of amylase enzyme.	
Sample Method 1. Put a test tube containing starch solution and a test tube containing amylase into a water bath at 37°C. 2. After 5 minutes add the amylase solution to the starch. 3. Every 30 seconds take a drop from the mixture and test it for starch using iodine solution. 4. Record how long it takes for the starch to be completely digested. 5. Repeat the experiment at different pH values using different buffer solutions.	**Considerations, Mistakes and Errors** • The solutions need to be left in the water bath for a while to reach the correct temperature before they are mixed. • After mixing, the tube must be kept in the water bath. • A buffer solution must be used to keep the reaction mixture at a certain fixed pH.
Variables • The independent variable is the one deliberately changed – in this case, the pH. • The dependent variable is the one that is measured – in this case, the time taken for the starch to be digested. • The control variables are kept the same – in this case, temperature, concentration and volume of starch and amylase.	**Hazards and Risks** • Care must be taken if a Bunsen is used to heat the water bath. • Take care not to spill iodine solution on the skin.

Key Point

Bile does not contain any enzymes, so it does not digest fat molecules. It just breaks up fat droplets into smaller ones.

Quick Test

1. What type of molecule is an enzyme?
2. Give **two** factors that affect the rate at which enzymes work.
3. Where are protease enzymes produced in the body?
4. What type of enzyme breaks down lipids?
5. Where is bile produced?

Key Words

catalyst
active site
optimum
lock and key theory
denature
protease
lipase
carbohydrase
amylase
bile

Blood and the Circulation

You must be able to:

- Explain how the components of blood are adapted for their roles
- Explain how the different types of blood vessels are adapted for their functions
- Describe the structure and function of the heart
- Explain the adaptations that allow gaseous exchange to take place between the lungs and the blood.

Blood

- Blood is a tissue.
- It is made of a liquid called **plasma**, which has three different components suspended in it:
  - red blood cells
  - white blood cells
  - platelets.
- Plasma transports various chemical substances around the body, such as the products of digestion, hormones, antibodies, urea and carbon dioxide.
- Red blood cells:
  - contain **haemoglobin**, which binds to oxygen to transport it from the lungs to the tissues and cells, which need it for respiration
  - do not contain a nucleus, so there is more room for haemoglobin
  - are very small, so they can fit through the tiny capillaries
  - are shaped like biconcave discs, giving them a large surface area that oxygen can quickly diffuse across.
- White blood cells:
  - help to protect the body against infection
  - can change shape, so they can squeeze out of the blood vessels into the tissues or surround and engulf microorganisms.
- Platelets are fragments of cells, which collect at wounds and trigger blood clotting.

Blood Vessels

- The blood passes around the body in blood vessels.
- The body contains three different types of blood vessel:

Arteries	Veins	Capillaries
<ul><li>Take blood from your heart to your organs.</li><li>Thick walls made from muscle and elastic fibres.</li></ul>	<ul><li>Take blood from your organs to your heart.</li><li>Thinner walls and valves to prevent backflow.</li></ul>	<ul><li>Allow substances needed by the cells to pass out of the blood.</li><li>Allow substances produced by the cells to pass into the blood.</li><li>Narrow, thin-walled blood vessels.</li></ul>

> ### Key Point
>
> In the lungs:
>
> haemoglobin + oxygen → oxyhaemoglobin
>
> In the tissues:
>
> oxyhaemoglobin → haemoglobin + oxygen

Red Blood Cell

No nucleus, so packed full of haemoglobin to absorb oxygen.

White Blood Cell

Can change shape in order to engulf and destroy invading microorganisms.

The Heart

- The heart pumps blood around the body in a **double circulatory system**.
- Blood passes through the heart twice on each circuit.
- There are four chambers in the heart:
  - the left and right **atria**, which receive blood from veins
  - the left and right **ventricles**, which pump the blood out into arteries.

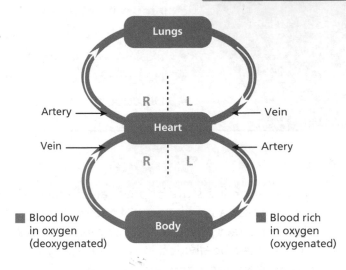

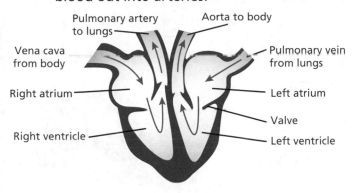

- Blood enters the heart through the atria.
- The atria contract and force blood into the ventricles.
- The ventricles then contract and force blood out of the heart.
- Valves make sure the blood flows in the correct direction.
- The natural resting heart rate is controlled by a group of cells located in the right atrium, which act as a **pacemaker**.
- Artificial pacemakers are electrical devices used to correct irregularities in the heart rate.

Gaseous Exchange

- The heart sends blood to the lungs via the **pulmonary artery**.
- Air obtained by breathing reaches the lungs through the **trachea** (windpipe).
- The trachea divides into two tubes – the **bronchi**.
- The bronchi divide to form **bronchioles**.
- The bronchioles divide until they end in tiny air sacs called **alveoli**.
- There are millions of alveoli and they are adapted to be very efficient at exchanging oxygen and carbon dioxide:
  - They have a large, moist surface area.
  - They have a very rich blood supply.
  - They are very close to the blood capillaries, so the distance for gases to diffuse is small.
- The blood is taken back to the heart through the **pulmonary vein**.

> ### Key Point
>
> The pulmonary artery is unusual because, unlike other arteries, it carries deoxygenated blood. The pulmonary vein carries oxygenated blood.

> ### Key Words
>
> plasma
> haemoglobin
> **double circulatory system**
> **atria**
> **ventricles**
> **pacemaker**
> **pulmonary artery**
> **trachea**
> **bronchi**
> **bronchioles**
> **alveoli**
> **pulmonary vein**

Quick Test

1. Which component of blood makes it clot?
2. How are red blood cells adapted to carry oxygen?
3. Which type of blood vessel carries blood away from the heart?
4. In which chamber does deoxygenated blood enter the heart?
5. What do the heart and veins contain to prevent backflow of blood?

Non-Communicable Diseases

You must be able to:

- Explain the difference between non-communicable and communicable diseases
- Explain, with examples, what is meant by 'risk factors'
- Describe the causes and treatments for certain heart diseases
- Describe the main causes and types of cancer.

Health and Disease

- Good **health** is a state of physical and mental wellbeing.
- A **disease** is caused by part of the body not working properly. This can affect physical and / or mental health.
- Diseases can be divided into two main types: **communicable diseases** and **non-communicable disease**.
- Non-communicable diseases cannot be spread between organisms, but communicable diseases can.
- There are many examples of how different diseases can interact with each other:
  - Viruses infecting cells can be the trigger for cancers, such as cervical cancer.
  - Diseases of the immune system mean that an individual is more likely to catch infectious diseases, e.g. people with HIV are more likely to get tuberculosis.
  - Immune reactions triggered by a pathogen can cause allergies, such as skin rashes and asthma.
  - If a person is physically ill, this can lead to depression and mental illness.
  - Poor diet, stress and difficult life situations can increase the likelihood of developing certain diseases.
- Non-communicable diseases, such as HIV and diabetes, can change a person's life and cost countries large sums of money.
- About 10% of the health budget in Britain is spent on people with diabetes.

Key Point

A causal mechanism is the process by which a cause brings about an effect.

A causal mechanism has been found that links smoking to lung cancer. It is the action of the chemicals in the tar.

Risk Factors

- Non-communicable diseases are often caused by the interaction of a number of factors.
- These factors are called **risk factors**, because they make it more likely that a person will develop the disease.
- Risk factors can be:
  - aspects of a person's lifestyle, e.g. lack of exercise
  - substances in the person's body or environment, e.g. chemicals from smoking.
- Sometimes there is a clear link between a risk factor and the chance of getting a disease, e.g. obesity and Type 2 diabetes.
- This does not necessarily mean that the risk factor causes the disease.
- Scientists need to look for a **causal mechanism** to prove that a risk factor is involved.

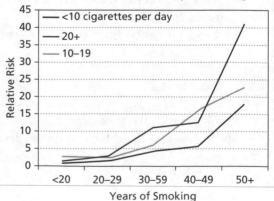

Relative Risk of Lung Cancer in Men According to Duration and Intensity of Smoking

- Causal mechanisms have been found, linking some diseases and risk factors.

Disease	Proven Risk Factors
Cardiovascular disease	Lack of exercise / smoking / high intake of saturated fat
Type 2 diabetes	Obesity
Liver and brain damage	Excessive alcohol intake
Lung diseases, including lung cancer	Smoking
Skin cancer	Ionising radiation, e.g. UV light
Low birth weight in babies	Smoking during pregnancy
Brain damage in babies	Excessive alcohol intake during pregnancy

Diseases of the Heart

- In **coronary heart disease**, layers of fatty material build up inside the coronary arteries and narrow them.
- Treatments for coronary heart disease include:
  - **stents** to keep the coronary arteries open
  - **statins** to reduce blood cholesterol levels and slow down the rate at which fatty materials build up.
- In some people, heart valves may become faulty, developing a leak or preventing the valve from opening fully.
- Faulty valves can be replaced using biological or mechanical valves.
- For cases of heart failure:
  - a donor heart, or heart and lungs, can be transplanted
  - artificial hearts can be used to keep patients alive while waiting for a heart transplant or to allow the heart to recover.
- Drugs such as clot-busting enzymes or warfarin are sometimes used to treat recovering patients, while statins can be given to lower cholesterol levels.

Cancer

- Cancer is a non-communicable disease.
- Scientists have identified lifestyle risk factors for some types of cancer, e.g. smoking, obesity, common viruses and UV exposure.
- There are also genetic risk factors for some cancers, which may run in families, e.g. some genes make the carrier more susceptible to certain types of breast cancer.
- Cancer is caused by uncontrolled cell division. This can form masses of cells called **tumours**.
- There are two main types of tumours:
  - **Benign** tumours do not spread around the body.
  - **Malignant** tumours spread, in the blood, to different parts of the body where they form secondary tumours.

Coronary Heart Disease

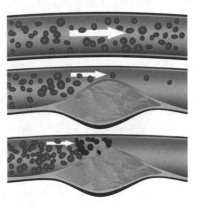

Key Point

The narrowing of coronary arteries reduces the flow of blood, so not enough oxygen can reach the heart muscle.

Key Words

health
disease
communicable
non-communicable
risk factor
causal mechanism
coronary heart disease
stent
statins
tumours
benign
malignant

Quick Test

1. Are each of these diseases non-communicable or communicable?
   a) Flu   b) Scurvy (lack of vitamin C)   c) Lung cancer
2. Why can spending too long in the sun result in skin cancer?
3. Give **one** risk factor for Type 2 diabetes.
4. Describe how coronary heart disease can cause heart muscle cells to stop contracting.

Transport in Plants

You must be able to:

- Explain how various plant tissues are adapted for their functions
- Describe how water is transported through a plant
- Describe how dissolved food substances are transported in a plant.

Plant Tissues

Tissue	Function
Epidermis	Covers the outer surfaces of the plant for protection.
Palisade mesophyll	The main site of photosynthesis in the leaf.
Spongy mesophyll	Air spaces between the cells allow gases to diffuse through the leaf.
Xylem vessels	Transports water and minerals through the plant, from roots to leaves. Also supports the plant.
Phloem vessels	Transports dissolved food materials through the plant.
Meristem tissue	Found mainly at the tips of the roots and shoots, where it can produce new cells for growth.

- Plant tissues are gathered together to form organs.
- The leaf is a plant organ.
- The structures of tissues in the leaf are related to their functions:

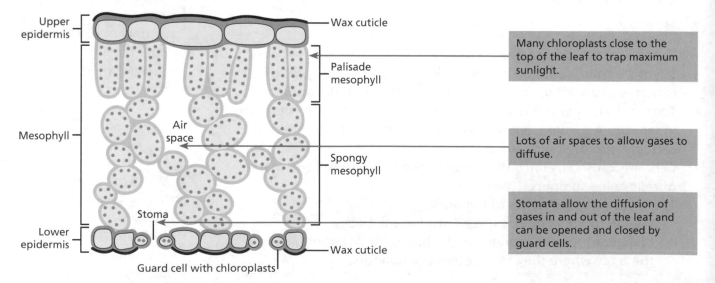

Water Transport

- Water enters the plant from the soil, through the root hair cells, by osmosis (see pages 14–15).
- Root hair, xylem and phloem cells are specialised to transport water, minerals and sugars around the plant (see page 16).

- This water contains dissolved minerals.
- The water and minerals are transported up the xylem vessels, from the roots to the stems and leaves.
- At the leaves, most of the water will evaporate and diffuse out of the **stomata** (small pores).
- The loss of water from the leaves is called **transpiration**.
- It helps to draw water up the xylem vessels from the roots.
- There are many factors that can affect the rate of transpiration:
  - An increase in temperature will increase the rate, as more energy is transferred to the water to allow it to evaporate.
  - Faster air flow will increase the rate, as it will blow away water vapour allowing more to evaporate.
  - Increased light intensity will increase the rate, as it will cause stomata to open.
  - An increase in humidity will decrease the rate as the air contains more water vapour, so the concentration gradient for diffusion is lower.
- In the leaf, the role of guard cells is to open and close stomata.
- At night the stomata are closed. This is because carbon dioxide is not needed for photosynthesis, so closing the stomata reduces water loss.
- When water is plentiful, guard cells take up water and bend. This causes the stomata to open, so gases for photosynthesis are free to move in and out of the stomata along with water from transpiration.
- When water is scarce, losing water makes the stomata change shape and close. This stops the plant from losing more water through transpiration.

Key Point

In most plants, the stomata are mainly found on the bottom surface of the leaf. This means that the sun does not shine directly on them, reducing water loss.

Measuring Water Uptake by a Leafy Shoot Using a Potometer

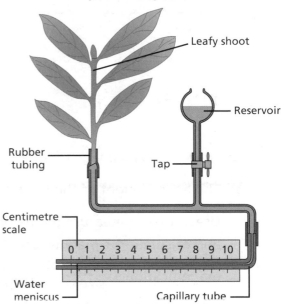

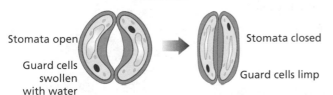

Guard cells

Stomata open — Guard cells swollen with water → Stomata closed — Guard cells limp

- The rate of transpiration from a cut shoot can be estimated by measuring the rate at which the shoot takes up water.
- This is only an estimate because not all of the water taken up by a shoot is lost – a very small percentage is used in the leaf.

Translocation

- Phloem tissue transports dissolved sugars from the leaves to the rest of the plant.
- This movement of food through phloem tissue is called **translocation**.
- Phloem cells are adapted for this function (see page 16).

Key Point

Plants cannot stop transpiration completely. This is because carbon dioxide is needed for photosynthesis, so water will always escape.

(see page 16)

Quick Test

1. In plant leaves, which tissue is the main site of photosynthesis?
2. What substances are transported by xylem tissue?
3. Give **two** environmental factors that slow down transpiration.
4. Why do stomata close at night?

Key Words

stomata
transpiration
translocation

Cell Structure

1. Complete **Table 1** with a tick (✓) or cross (✗) to show if the structures are present or absent in the cells listed.

Table 1

Type of Cell	Nucleus	Cytoplasm	Cell Membrane	Cell Wall
Plant cell	✓	✓	✓	✓
Bacterial cell	✗	✓	✓	✗
Animal cell	✓	✓	✓	✗

[3]

Total Marks _____ / 3

Investigating Cells

1. **Figure 1** shows an image of a palisade cell.

   a) The actual height of the cell is 0.1mm.

   Use a ruler to measure the height of the cell in the diagram and calculate the magnification of the image in **Figure 1**.

   $$\text{magnification} = \frac{\text{size of image}}{\text{size of real object}}$$

   [2]

   b) Which structure, **A, B, C, D** or **E**, in the cell in **Figure 1** matches each of the following descriptions?

   You may use the same letter more than once.

   i)   Where most chemical reactions take place. E [1]

   ii) Made of cellulose. A [1]

   iii) Contains chlorophyll. C [1]

   iv) Controls the cell's functions. B [1]

   v)   Gives the cell rigidity and strength. A [1]

   vi) Filled with cell sap. D [1]

   vii) Where photosynthesis occurs. C [1]

   viii) Contains chromosomes. B [1]

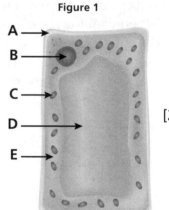

Figure 1

A →
B →
C →
D →
E →

2 There are many different types of microorganisms that live in soil.

Table 1 gives the average number of each type of microorganism in one gram of soil.

Table 1

Type of Microorganism	Number in One Gram of Soil
viruses	160 000 000
bacteria	3 000 000
fungi	1 100 000

a) HT What is the average number of bacteria in **10 grams** of soil?

Tick **one** box.

0.3×10^6 ☐ 3×10^6 ☐ 3×10^7 ☑ 13×10^6 ☐

[1]

b) Tilly wants to investigate the bacteria in soil from her garden.

This is the method she uses:

1. Mix one gram of soil with a small volume of water.

2. Spread the mixture on an agar plate and fix the lid on with a piece of tape.

3. Incubate the plate upside down at 25°C.

i) Give **two** reasons why Tilly must make sure that the lid of the Petri dish cannot come off. *So oxyyen or new bakeria doesr get in* [2]

ii) Why does Tilly incubate the dishes upside down? [1]
Stop condensation falling on bacteria

Tilly repeats this process using one gram of soil diluted down 1 000 000 times.

c) **Figure 2** shows Tilly's plates after they have been incubated.

Figure 2

i) Explain why individual colonies of bacteria cannot be seen growing on Tilly's first plate. [2]
too much bacteria

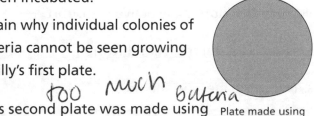

Plate made using first sample

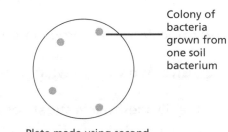

Colony of bacteria grown from one soil bacterium

Plate made using second (diluted) sample

ii) Tilly's second plate was made using one gram of soil that had been diluted down 1 000 000 times.

Does Tilly's soil contain more or less than the average number of bacteria? You must show how you have arrived at your answer. [3]

Total Marks .... / 19

Practice Questions

Cell Division

1 A magazine published this article about stem cells:

> In 2006, scientists at Kyoto University managed to turn body cells from rats into stem cells, by inserting some genes.
>
> These genes seemed to reprogramme the cells so that they could use all their genes again. This made the cells act like embryonic stem cells.
>
> Doctors are very interested in this discovery. This is because they hope that the technique could be used to produce stem cells in a way that fewer people would object to.

a) What are embryonic stem cells?   [2]

b) Explain why stem cells may be useful to doctors and why this discovery might make fewer people object to their use.   [5]

Total Marks _____ / 7

Transport In and Out of Cells

1 **a)** Complete the sentences about diffusion.   [4]

Diffusion is the spreading of _____ from an area of _____

concentration to an area of _____ concentration.

The greater the difference in concentration, the _____ the rate of diffusion.

b) Glucose is reabsorbed into the blood in the kidneys by active transport.

Give **two** ways in which active transport is different from diffusion.   [2]

2 Katie decides to investigate osmosis. This is the method she uses:

1. Prepare five boiling tubes containing different sugar solutions.
2. Cut five potato chips of equal size.
3. Weigh each potato chip and place one in each tube.
4. Leave for 24 hours.
5. Reweigh the potato chips.

Table 1 shows her results.

Table 1

Concentration of Sugar Solution (mol/dm³)	Mass of Potato Chip Before (g)	Mass of Potato Chip After (g)	Difference in Mass (g)
0.00	1.62	1.74	0.12
0.25	1.72	1.62	−0.10
0.50	1.69	1.62	−0.07
0.75	1.76	1.60	−0.16
1.00	1.74	1.59	−0.15

a) What is the independent variable in Katie's investigation? [1]

b) Name a variable that should be kept constant during Katie's investigation. [1]

c) In which concentration did the potato gain mass? [1]

d) Explain why this potato chip gained mass. [3]

e) How could Katie increase the reliability of her results? [1]

3 Complete **Table 2** by putting a tick (✓) or a cross (✗) in each of the blank boxes.

Table 2

	Osmosis	Diffusion	Active Transport
Can cause a substance to enter a cell		✓	
Needs energy from respiration	✗		
Can move a substance against a concentration gradient	✗		
Is responsible for oxygen moving into the red blood cells in the lungs			✗

[4]

Total Marks _____ / 17

Levels of Organisation

1   Some cells are specialised to carry out a specific function.

Draw **one** line from each description to the correct function of that type of cell.

Description	Function
A cell that is hollow and forms tubes	to contract
A cell that has a flagellum	to transport water
A cell that is full of protein fibres	to carry nerve impulses
A cell that has a long projection with branched endings	to swim

[3]

Total Marks _____ / 3

Digestion

1   a) Name the organ in the body that produces bile.   [1]

b) Where is bile stored in the human body?   [1]

c) Into which part of the digestive system is bile released?   [1]

d) Bile is needed to neutralise the acid that was added to food in the stomach.

Why is this important?   [1]

2   Enzymes are used in industry and in the home.
For example, enzymes can be used in the manufacture of baby food to help predigest certain foods.

a) i)   What type of enzyme is used in industry to predigest proteins?   [1]

ii) What are produced through the digestion of proteins by enzymes?   [1]

b) A baby food manufacturer wants to improve the efficiency of the business by using the enzyme that digests protein the fastest.

The manufacturer already uses enzyme **X**, which takes 15 minutes to completely digest the protein.

The manufacturer employs a scientist to investigate four other enzymes **A, B, C** and **D**.

In each trial, the same concentration of enzyme and mass of protein is used.

The time taken for each enzyme to completely digest the protein is recorded.

Table 1 shows the results.

Table 1

Enzyme	A	B	C	D
Time Taken to Digest Protein (Minutes)	19	6	13	16

i)   What is the independent variable in this investigation?   [1]

ii) Write down the step that the scientist takes to try to get valid results.   [1]

iii) The scientist started plotting the results on the graph in **Figure 1** below.

Complete the graph in **Figure 1** by plotting the remaining results.

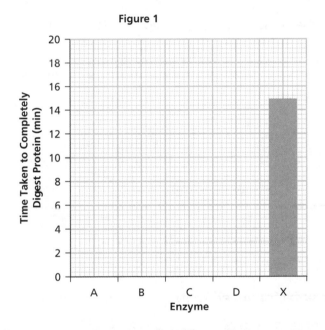

Figure 1

[4]

c) Which enzyme would you recommend the manufacturer should use?
Give a reason for your answer.   [2]

d) The research and development team at the company are not convinced by the results of
this investigation.

What could the scientist do to increase the reliability of the results?   [1]

Total Marks _____ / 15

Practice Questions

Blood and the Circulation

1 **Figure 1** shows some of the structures in the thorax.

Figure 1

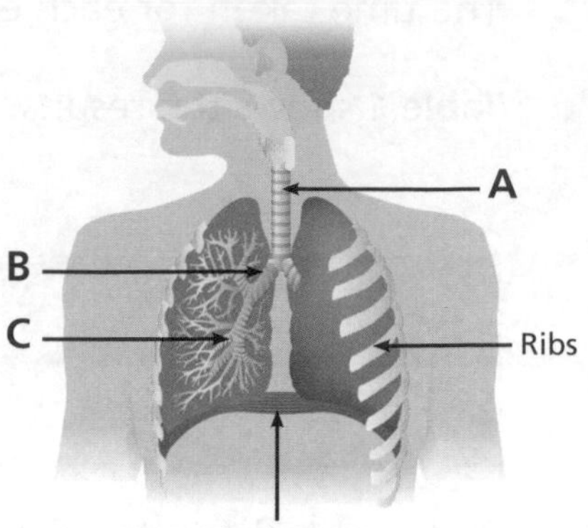

a) Name the structures **A**, **B** and **C**. [3]

b) Alveoli are adapted to help gas exchange in the lungs.

   i) Which gas diffuses from the blood into the alveoli? [1]

   ii) Which gas diffuses from the alveoli into the blood? [1]

2 The human body has a double circulatory system.

Complete the sentences.

Oxygenated blood from the _____ returns to the left atrium of the heart in the

pulmonary _____. From here, it enters the left ventricle and leaves the heart via

the aorta to go to the _____. From the body, _____ blood returns via

the _____ to the right atrium and then leaves the heart in the pulmonary

artery to go to the lungs. [5]

> **Total Marks** _____ / 10

Non-Communicable Diseases

1 Some non-communicable diseases have proven risk factors linked to them.

Draw **one** line from each disease to the risk factor linked to it.

Disease	Proven Risk Factors
Type 2 diabetes	smoking
liver damage	excess alcohol intake
lung cancer	obesity
skin cancer	UV light

[3]

2 Emphysema is a lung disease that irreversibly damages the alveoli in the lungs.

Two men do the same amount of daily exercise.

One man is in good health. The other man has emphysema.

Measurements are taken to show:
- The amount of air that enters their lungs when they inhale.
- The amount of oxygen that enters their blood.

Table 1 shows the results.

Table 1

	Healthy Man	**Man with Emphysema**
Total Air Flowing Into Lungs (dm³ per minute)	89.5	38.9
Oxygen Entering Blood (dm³ per minute)	2.5	1.2

a) Calculate the percentage difference between the total air flowing into the two men's lungs per minute. [2]

b) Explain how the changes to the lungs caused by emphysema can account for the difference in the oxygen figures. [3]

c) Explain why the man with emphysema will struggle to carry out exercise. [2]

Total Marks .... / 10

Transport in Plants

1 Circle **three** phrases that could be used to correctly complete the sentence.

absorbing oxygen   giving off water vapour   absorbing carbon dioxide

giving off carbon dioxide   giving off oxygen   absorbing water

During the day, when the stomata are open, the leaf is .... [3]

2 Plants have two separate transport tissues.

a) Give the name of the tissue responsible for transporting water and mineral ions. [1]

b) Give the name of the tissue responsible for transporting dissolved sugars. [1]

Total Marks .... / 5

Pathogens and Disease

You must be able to:

- Describe the main types of disease-causing pathogen
- Describe the symptoms and method of spread of measles, HIV, salmonella and gonorrhoea
- Explain the role of the mosquito in the spread of malaria
- Describe the symptoms and method of spread of rose black spot in plants.

Pathogens and Disease

- **Pathogens** are microorganisms that cause infectious (communicable) diseases.
- Pathogens may infect plants or animals.
- They can be spread by:
  - direct contact
  - water or air
  - **vectors** (organisms that carry and pass on the pathogen without getting the disease).
- The spread of infectious diseases can be reduced by:
  - simple hygiene measures, e.g. washing hands and sneezing into a handkerchief
  - destroying vectors
  - isolating infected individuals, so they cannot pass the pathogen on
  - giving people at risk a vaccination (see page 37).

Viral Pathogens

- Viruses reproduce rapidly in body cells, causing damage to the cells.
- **Measles** is a disease caused by a virus:
  - The symptoms are fever and a red skin rash.
  - The measles virus is spread by breathing in droplets from sneezes and coughs.
  - Although most people recover well from measles, it can be fatal if there are complications, so most young children are vaccinated against measles.
- **HIV** (human immunodeficiency virus) causes AIDS:
  - It is spread by sexual contact or exchange of body fluids, e.g. it can be transmitted in blood when drug users share needles.
  - At first, HIV causes a flu-like illness.
  - If untreated, the virus enters the lymph nodes and attacks the body's immune cells.
  - Taking antiviral drugs can delay this happening.
  - Late stage HIV, or AIDS, is when the body's immune system is damaged and cannot fight off other infections or cancers.
- Viruses can also cause plant diseases, for example tobacco mosaic virus (see page 40).

The Virus that Causes Measles

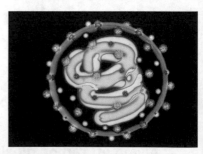

> ### Key Point
>
> It is not the HIV virus that directly kills people with AIDS. It is other infections, such as pneumonia, that a healthy body would usually be able to survive.

Bacterial Diseases

- Bacteria may damage cells directly or produce **toxins** (poisons) that damage tissues.
- **Salmonella** is a type of food poisoning caused by bacteria:
  - The bacteria are ingested in food, which may not have been cooked properly or may not have been prepared in hygienic conditions.
  - The bacteria secrete toxins, which cause fever, abdominal cramps, vomiting and diarrhoea.
  - Chicken and eggs can contain the bacteria, so chickens in the UK are vaccinated against salmonella to control the spread.
- **Gonorrhoea** is a sexually transmitted disease (STD) caused by bacteria:
  - It is spread by sexual contact.
  - The symptoms are a thick, yellow or green discharge from the vagina or penis and pain when urinating.
  - It used to be easily treated with penicillin, but many resistant strains have now appeared.
  - The use of a barrier method of contraception, e.g. a condom, can stop the bacteria being passed on.

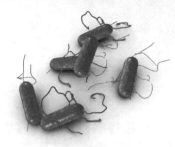

The Bacteria that Cause Salmonella

Protists and Disease

- Protists are single-celled organisms.
- However, unlike bacteria, they are eukaryotic.
- **Malaria** is caused by a protist:
  - The protist uses a particular type of mosquito as a vector.
  - It is passed on to a person when they are bitten by the mosquito.
  - Malaria causes severe fever, which reoccurs and can be fatal.
  - One of the main ways to stop the spread is to stop people being bitten, e.g. by killing the mosquitoes or using mosquito nets.

Mosquitoes Transmit Malaria

Fungal Diseases

- **Rose black spot** is a fungal disease:
  - It is spread when spores are carried from plant to plant by water or wind.
  - Purple or black spots develop on leaves, which often turn yellow and drop early.
  - The loss of leaves will stunt the growth of the plant because photosynthesis is reduced.
  - It can be treated by using fungicides and removing and destroying the affected leaves.

Quick Test

1. State one simple precaution that can stop pathogens being spread by droplets in the air.
2. Why is it important to keep uncooked meat separate from cooked meat?
3. Why does the contraceptive pill **not** prevent the spread of gonorrhoea?
4. Why should leaves infected with rose black spot be removed and burned?

Key Point

In malaria, the protist is the pathogen for the disease. The mosquito is acting as a parasite when it feeds on a person.

Key Words

pathogen
vector
toxin

Human Defences Against Disease

You must be able to:

- Describe how the body tries to prevent pathogens from entering
- Describe how the immune system reacts if pathogens do enter the body
- Explain the process of immunity and how vaccinations work.

Preventing Entry of Pathogens

- The body has a number of **non-specific defences** against disease.
- These are defences that work against all pathogens, to try and stop them entering the body.

The Body's Defences

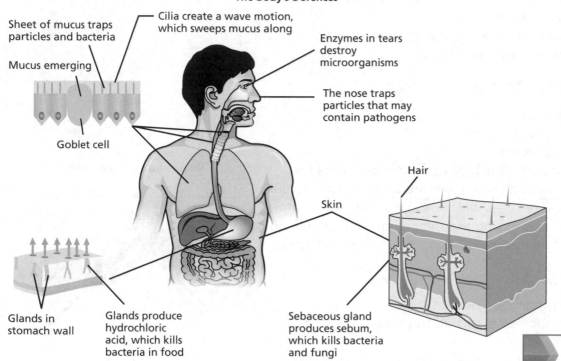

Sheet of mucus traps particles and bacteria

Mucus emerging

Goblet cell

Cilia create a wave motion, which sweeps mucus along

Enzymes in tears destroy microorganisms

The nose traps particles that may contain pathogens

Hair

Skin

Glands in stomach wall

Glands produce hydrochloric acid, which kills bacteria in food

Sebaceous gland produces sebum, which kills bacteria and fungi

The Immune System

- If a pathogen enters the body, the **immune system** tries to destroy it.
- White blood cells help to defend against pathogens through:
  - **phagocytosis**, which involves the pathogen being surrounded, engulfed and digested

> ### Key Point
>
> Antibodies are specific to a particular pathogen, e.g. antibodies against gonorrhoea bacteria will not attach to salmonella bacteria.

White blood cell

Microorganisms invade the body.

The white blood cell finds the microorganisms and engulfs them.

The white blood cell ingests the microorganisms.

The microorganisms have been digested and destroyed.

- the production of special protein molecules called **antibodies**, which attach to **antigen** molecules on the pathogen

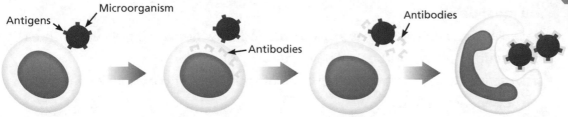

Antigens — Microorganism

Antibodies

Antibodies

Antibodies

| Antigens are markers on the surface of the microorganism. | The white blood cells become sensitised to the antigens and produce antibodies. | The antibodies then lock onto the antigens. | This causes the microorganisms to clump together, so that other white blood cells can digest them. |

- the production of **antitoxins**, which are chemicals that neutralise the poisonous effects of the toxins.

Boosting Immunity

- If the same pathogen re-enters the body, the white blood cells respond more quickly to produce the correct antibodies.
- This quick response prevents the person from getting ill and is called **immunity**.
- When a person has a **vaccination**, small quantities of dead or inactive forms of a pathogen are injected into the body.
- Vaccination stimulates the white blood cells to produce antibodies and to develop immunity.

1 A weakened / dead strain of the microorganism is injected. Antigens on the modified microorganism's surface cause the white blood cells to produce specific antibodies.

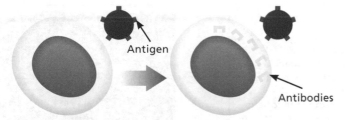

Antigen

Antibodies

2 The white blood cells that are capable of quickly producing the specific antibody remain in the bloodstream.

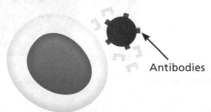

Antibodies

- If a large proportion of the population can be made immune to a pathogen, then the pathogen cannot spread very easily.

> ### Key Point
>
> The disease smallpox does not exist anymore. This is because vaccination managed to prevent the pathogen spreading to new hosts. Scientists are hoping to repeat this with polio.

> ### Key Point
>
> Some vaccinations do carry a very small risk of side effects, but it is important to compare this with the risk of getting the disease.

> ### Key Words
>
> **non-specific defences**
> **immune system**
> **phagocytosis**
> **antibody**
> **antigen**
> **antitoxin**
> **immunity**
> **vaccination**

Quick Test

1. How does the stomach help to kill pathogens?
2. What is phagocytosis?
3. What is the name of the protein molecules made by white blood cells when they detect a pathogen?
4. Why does vaccination use a dead or weakened pathogen?

Treating Diseases

You must be able to:

- Explain how antibiotics have saved lives and why their use is now under threat
- Describe how new drugs are developed
- HT Describe the production of monoclonal antibodies and explain why they are useful.

Antibiotics

- **Antibiotics**, e.g. penicillin, are medicines that kill bacteria inside the body. However, they cannot destroy viruses.
- Doctors will prescribe certain antibiotics for certain diseases.
- The use of antibiotics has greatly reduced deaths from infections.
- However, bacterial strains resistant to antibiotics are increasing (see page 80).
- **MRSA** is a strain of bacteria that is resistant to antibiotics.
- To reduce the rate at which resistant strains of bacteria develop:
  - doctors should **not** prescribe antibiotics:
    - o unless they are really needed
    - o for non-serious infections
    - o for viral infections.
  - patients must complete their course of antibiotics so that all bacteria are killed and none survive to form resistant strains.

REQUIRED PRACTICAL	
Investigating the effect of different antibiotics on bacterial growth.	
Sample Method 1. Inoculate a Petri dish with a culture of bacteria (see page 11). 2. Soak small discs of filter paper in different antibiotics. 3. Using forceps, place the antibiotic discs on the surface of the agar. 4. Incubate the sealed dish upside down at 25°C for several days.	**Considerations, Mistakes and Errors** • It is also possible to vary the concentration of the antibiotic to find the best concentration to use.
Variables • The independent variable is the type of antibiotic. • The dependent variable is the area around each disc that is clear of bacteria. The greater the area, the more effective the antibiotic is in killing the bacteria. • The control variables are the concentration of antibiotic used and the length of time that the discs are soaked.	**Hazards and Risks** • Care must be taken to follow aseptic techniques (see page 11).

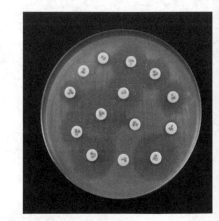

Developing New Drugs

- There is a constant demand to produce new drugs:
  - New painkillers are developed to treat the symptoms of disease but they do not kill the pathogens.
  - Antiviral drugs are needed that will kill viruses without also damaging the body's tissues.
  - New antibiotics are needed as resistant strains of bacteria develop.

- Traditionally drugs were extracted from plants and microorganisms:
  - **Digitalis** is a heart drug that originates from foxgloves.
  - **Aspirin** is a pain killer that originates from willow.
  - **Penicillin** was discovered by Alexander Fleming from the *Penicillium* mould.
- Now most new drugs are synthesised (made) by chemists in the pharmaceutical industry. However, the starting point may still be a chemical extracted from a plant.
- New medical drugs have to be tested and trialled before being used to make sure they are safe (not toxic).
- If a drug is found to be safe, it is then tested on patients to:
  - see if it works
  - find out the optimum dose.
- These tests on patients are usually **double-blind trials**:
  - some patients are given a **placebo**, which does not contain the drug, and some patients are given the drug
  - patients are allocated randomly to the two groups
  - neither the doctors nor the patients know who has received a placebo and who has received the drug.
- New painkillers are developed to treat the symptoms of disease – they do not kill pathogens
- New antiviral drugs are needed that will kill viruses without damaging the body's tissues. This is not easy to achieve.

Monoclonal Antibodies

- **Monoclonal antibodies** are produced from a single cell that has divided to make many cloned copies of itself.
- These antibodies bind to only one type of antigen, so they can be used to target a specific chemical or specific cells in the body.
- They are produced by combining mouse cells and a tumour cell to make a cell called a **hybridoma**.
- Monoclonal antibodies can be used in different ways:
  - in pregnancy tests, to bind to the hormone **HCG** found in urine during early pregnancy
  - in laboratories, to measure the levels of hormones and other chemicals in blood, or to detect pathogens
  - in research, to locate or identify specific molecules in a cell or tissue by binding to them with a fluorescent dye
  - to treat some diseases, e.g. in cancer they can be used to deliver a radioactive substance, a toxic drug, or a chemical that stops cells dividing, specifically to the cancer cells.
- Unfortunately, monoclonal antibodies have created more side effects than expected, so they are not yet widely used.

> ### Key Point
>
> The purpose of a double-blind test is to ensure that it is completely fair. If the patients or doctors knew whether it was the drug or a placebo being used, it might influence the outcome of the test.

HT **Monoclonal Antibodies**

Vaccinate mouse to stimulate the production of antibodies

Collect spleen cells that form antibodies from mouse
Tumour cells (myeloma)

Spleen and myeloma cells fuse to form hybridoma cells

Grow hybridoma cells in tissue culture and select antibody-forming cells

Collect monoclonal antibodies

> ### Key Words
>
> antibiotics
> MRSA
> digitalis
> aspirin
> penicillin
> double-blind trial
> placebo
> HT **monoclonal antibody**
> HT **hybridoma**
> HT **HCG**

> ### Quick Test
>
> 1. Why can antibiotics not be used to destroy HIV?
> 2. What term describes bacterial pathogens that are not affected by antibiotics?
> 3. Why did people once chew willow bark if they had a headache?
> 4. Why does the title 'double-blind trial' include the word 'double'?

Plant Disease

You must be able to:

HT Describe how plant diseases can be detected and identified
- Describe the cause and symptoms of certain plant diseases
- Describe various methods that plants use to defend themselves from disease.

HT Detecting and Identifying Plant Disease

- There are a number of signs that a plant may be diseased:
    - stunted growth
    - spots on leaves
    - areas of decay (rot)
    - growths
    - malformed (abnormal) stems or leaves
    - discolouration
    - the presence of pests.
- To identify the disease, a number of steps can be taken:
    - consulting a gardening manual or website
    - taking infected plants to a laboratory to identify the pathogen
    - using testing kits, which contain monoclonal antibodies.

Examples of Plant Diseases

- Like animals, plants can suffer from non-communicable and communicable diseases.
- Plants can be infected by a range of viral, bacterial and fungal pathogens as well as by insects.
- **Tobacco mosaic virus** (TMV) is a widespread plant pathogen:
    - It infects tobacco plants and many other plants, including tomatoes.

- It produces a distinctive 'mosaic' pattern of discolouration on the leaves, which reduces chlorophyll content and affects photosynthesis.
  - It affects the growth of the plant due to lack of photosynthesis.
- Rose black spot is a fungal disease (see page 35).
- **Aphids** are small insects often known as greenfly or blackfly. They feed from the phloem, taking sugars away from the plant.
- Non-communicable diseases include a range of **deficiency diseases**, caused by a lack of mineral ions in the soil:
  - Stunted growth is caused by nitrate deficiency, because nitrates are needed for protein synthesis.
  - **Chlorosis** is caused by magnesium deficiency, because magnesium ions are needed to make chlorophyll.

Plant Defences

- Plants have a number of physical defences to try and stop organisms entering them:
  - cellulose cell walls
  - a tough waxy cuticle on leaves
  - layers of dead cells around stems (bark on trees), which fall off and take pathogens with them.
- Some plants also have chemical defences, such as:
  - antibacterial chemicals, which are made by plants such as mint and witch hazel
  - poisons to deter herbivores, which are made by plants such as tobacco, foxgloves and deadly nightshade.
- Other plants have evolved with mechanical adaptations, including:
  - thorns and hairs to deter animals from eating or touching them
  - leaves that droop or curl when touched
  - **mimicry** to trick animals into not eating them or not laying eggs on the leaves, e.g. the white deadnettle does not sting, but it looks very similar to a stinging nettle.

Mimosa Leaves Curl Up to Deter Insects from Eating Them

White Deadnettle

Quick Test

1. Explain why the change in leaf colour caused by TMV can reduce photosynthesis.
2. What do aphids feed on?
3. Why does a lack of nitrates in the soil cause stunted growth?

Key Words

tobacco-mosaic virus (TMV)
aphid
deficiency disease
chlorosis
mimicry

 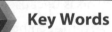

Photosynthesis

You must be able to:

- Recall the word equation for photosynthesis
- Understand that photosynthesis is an endothermic reaction
- Explain how various factors can change the rate of photosynthesis
- Describe how the products of photosynthesis are used by plants.

Photosynthesis

- The equation for photosynthesis is:

LEARN

$$\text{carbon dioxide} + \text{water} \xrightarrow{\text{light}} \text{glucose} + \text{oxygen}$$

$$CO_2 \qquad\qquad H_2O \qquad\qquad C_6H_{12}O_6 \qquad O_2$$

- To produce glucose molecules by photosynthesis, energy is required.
- This is because the reactions are **endothermic** (take heat in).
- The energy needed is supplied by sunlight.
- It is trapped by the green chemical **chlorophyll**, which is found in chloroplasts.

Factors Affecting Photosynthesis

- There are several factors that may affect the rate of photosynthesis.

HT At any moment, the factor that stops the reaction going any faster is called the **limiting factor**.

- **Temperature**
  - As the temperature increases, so does the rate of photosynthesis.
  - This is because more energy is provided for the reaction.
  - As the temperature approaches 45°C, the rate of photosynthesis drops to zero because the enzymes controlling photosynthesis have been destroyed.

- **Carbon dioxide concentration**
  - As the concentration of CO_2 increases, so does the rate of photosynthesis.
  - This is because CO_2 is needed in the reaction.
  - **HT** After reaching a certain point, an increase in CO_2 has no further effect. CO_2 is no longer the limiting factor.

- **Light intensity**
  - As light intensity increases, so does the rate of photosynthesis.
  - This is because more energy is provided for the reaction.
  - **HT** After reaching a certain point, any increase in light has no further effect. It is no longer the limiting factor.

- **Chlorophyll concentration**
  - This does not vary in the short term but may change if plants are grown in soil without enough minerals to make chlorophyll.

> ### Key Point
>
> During photosynthesis, the energy from sunlight is converted to chemical energy in the form of glucose molecules.

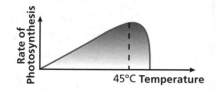

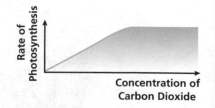

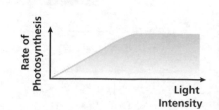

HT By looking at a graph, it can be possible to say what the limiting factor is at any point.

HT Greenhouses can be used to increase the rate of photosynthesis. By controlling lighting, temperature and carbon dioxide, farmers can increase the growth rate of their crops.

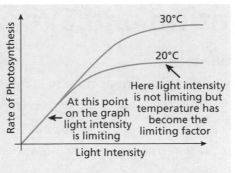

REQUIRED PRACTICAL	
Investigating the effect of a factor on the rate of photosynthesis.	
Sample Method One of the most common ways of measuring photosynthesis involves observing oxygen output from a piece of pondweed: 1. Place a piece of pondweed in a beaker and shine a light at it using a lamp a specific distance away. 2. Record the number of bubbles of gas coming out of the pondweed in one minute. 3. Repeat this with the lamp at different distances from the pond weed.	**Considerations, Mistakes and Errors** • It is best to take at least two readings at each distance and calculate the mean of the number of bubbles. • Carbon dioxide is provided by adding a small amount of sodium hydrogen carbonate to the water.
Variables • The independent variable is the light intensity (distance from the light). • The dependent variable is the number of bubbles in one minute. • The control variables are the piece of pondweed, the temperature, and the concentration of carbon dioxide.	**Hazards and risks** • Care must be taken to avoid any water being dropped onto the hot light bulb.

HT When light intensity is studied, doubling the distance between the lamp and the pondweed will reduce the light intensity by a quarter. This is called the **inverse square law**.

Converting Glucose

- The glucose produced in photosynthesis may be used by the plant during respiration to provide energy.
- Glucose may also be changed into other products such as:
  - insoluble starch, which is stored in the stem, leaves or roots
  - fat or oil, which is also stored, e.g. in seeds
  - cellulose, to strengthen cell walls
  - proteins, which are used for growth and for enzymes.
- To produce proteins from glucose, plants also use nitrate ions, which are absorbed from the soil.

Quick Test

1. Name the green pigment essential for photosynthesis.
2. Where do plants obtain the carbon dioxide used in photosynthesis?
3. List three factors that may limit the rate of photosynthesis.
4. What do plants need, in addition to glucose, to make proteins?

 Key Point

Farmers have to carefully work out if the extra cost of lighting and heating will be paid for by the extra growth that their crops achieve.

Key Point

Nitrate ions are needed to make proteins because amino acids contain nitrogen, but glucose does not.

Key Words

endothermic
chlorophyll
HT limiting factor
HT inverse square law

Respiration and Exercise

You must be able to:

- Explain why respiration is important in living organisms
- Compare the processes of aerobic and anaerobic respiration
- Explain the changes that occur in respiration during exercise
- Describe respiration as a part of the metabolism of the body.

The Importance of Respiration

- Respiration is an example of an **exothermic** reaction.
- It releases energy from glucose molecules for use by the body.
- Organisms need this energy:
  - for chemical reactions to build larger molecules
  - for movement
  - to keep warm.
- Respiration in cells can be **aerobic** (with oxygen) or **anaerobic** (without oxygen).

Aerobic Respiration

- The equation for aerobic respiration is the same in all organisms:

LEARN

glucose + oxygen $\longrightarrow$ carbon dioxide + water

$C_6H_{12}O_6$ $\quad$ O_2 $\qquad$ CO_2 $\qquad$ H_2O

Anaerobic Respiration

- In anaerobic respiration, the glucose is not completely broken down.
- This means that it transfers much less energy than aerobic respiration.
- The process of anaerobic respiration is different in animals to the process found in plants and yeast.
- In animals, lactic acid is produced:

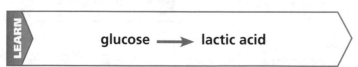

LEARN

glucose $\longrightarrow$ lactic acid

- In plants and yeast, alcohol (ethanol) and carbon dioxide are produced:

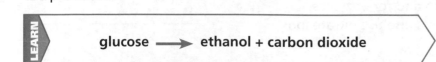

LEARN

glucose $\longrightarrow$ ethanol + carbon dioxide

- Anaerobic respiration in yeast cells is called **fermentation**.
- It is important in the manufacture of bread and alcoholic drinks.

Exercise and Respiration

- During exercise, the body demands more energy, so the rate of respiration needs to increase.

Key Point

Ethanol is the type of alcohol made by fermentation and is found in alcoholic drinks. The carbon dioxide produced can also be trapped to make the drink fizzy.

- The heart rate, breathing rate and breath volume all increase to supply the muscles with more oxygen and glucose for the increase in aerobic respiration.

Diffusion Between a Capillary and a Working Muscle Cell

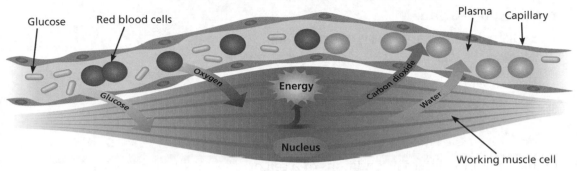

- During periods of vigorous activity, the muscles may not get supplied with enough oxygen, so anaerobic respiration starts to take place in the muscle cells.
- This causes a build-up of **lactic acid** and creates an **oxygen debt**.
- The lactic acid causes the muscles to hurt and stops them contracting efficiently. Lactic acid is a poison, so needs to be got rid of quickly.
- Once exercise is finished, the oxygen debt must be 'repaid'.

HT After exercise, blood flowing through the muscles transports the lactic acid to the liver where it is broken down.

HT The oxygen debt is the amount of extra oxygen the body needs after exercise to react with the lactic acid and remove it from the cells.

> **Key Point**
>
> Deep breathing for some time after exercise is used to 'pay back' the oxygen debt.

Metabolism

- **Metabolism** is the sum of all the chemical reactions in a cell or in the body.
- These reactions are controlled by enzymes and many need a transfer of energy.
- This energy is transferred by respiration and used to make new molecules.
- This includes:
  - the conversion of glucose to starch, glycogen and cellulose
  - the formation of lipid molecules from a molecule of glycerol and three molecules of fatty acids
  - the use of glucose and nitrate ions to form amino acids, which are used to synthesise proteins
  - the breakdown of excess proteins into urea for excretion.

> **Key Words**
>
> exothermic
> aerobic respiration
> anaerobic respiration
> fermentation
> lactic acid
> oxygen debt
> metabolism

> **Quick Test**
>
> 1. Which type of respiration requires oxygen?
> 2. Why do we need to eat more in cold weather?
> 3. Name the waste product of anaerobic respiration in animals
> 4. HT Where in the body is lactic acid broken down after exercise?
> 5. Why might your muscles hurt if you run a long race?

Homeostasis and Body Temperature

You must be able to:

- Explain why homeostasis is so important
- HT Describe the process of negative feedback
- Describe how body temperature is controlled.

The Importance of Homeostasis

- **Homeostasis** is the regulation of the internal conditions of a cell or organism in response to internal and external changes.
- Homeostasis is important because it keeps conditions constant for enzyme action and cell functions.
- Homeostasis includes the control of:
  - blood glucose concentration
  - body temperature
  - water and ion levels.
- The control systems may involve:
  - responses using nerves
  - chemical responses using hormones.

Control Systems

- All control systems include:
  - cells called **receptors**, which detect stimuli (changes in the environment)
  - coordination centres (such as the brain, spinal cord and pancreas), which receive and process information from receptors
  - **effectors** (muscles or glands), which bring about responses that restore optimum levels.

HT This type of control mechanism is called **negative feedback**:

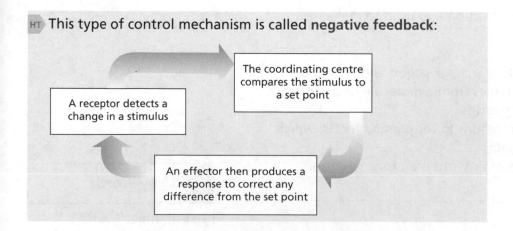

A receptor detects a change in a stimulus

The coordinating centre compares the stimulus to a set point

An effector then produces a response to correct any difference from the set point

Control of Body Temperature

- Human body temperature should be kept at around 37°C.
- This gives an optimum temperature for enzymes to work.
- The **thermoregulatory centre** in the brain:
  - monitors and controls body temperature

- has receptors that monitor the temperature of the blood flowing through the brain
- receives information (impulses) from temperature receptors in the skin.

- If the body temperature is too high:
  - blood vessels widen, directing more blood to the surface of the skin – this is called **vasodilation**
  - more sweat is produced from the sweat glands and evaporates
  - both these mechanisms cause a transfer of energy from the skin to the environment.

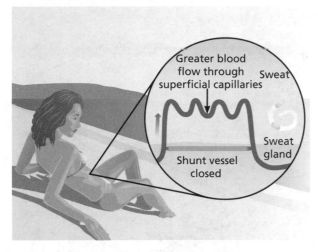

- If the body temperature is too low:
  - blood vessels narrow, directing blood away from the surface of the skin – this is called **vasoconstriction**
  - sweating stops
  - skeletal muscles uncontrollably contract and relax quickly (shiver), which transfers more heat to the blood.

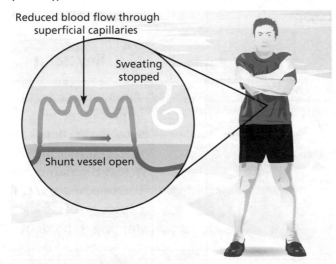

Key Point

The temperature receptors in the skin give the brain an early warning of changes in the external temperature. This allows behaviour to be adjusted to stop the body temperature changing, e.g. putting on more clothes.

Key Point

Blood vessels in the skin cannot move, so narrowing or widening is the only way that they can adjust how much energy is transferred to the outside.

Key Words

homeostasis
receptors
effectors
HT negative feedback
thermoregulatory centre
vasodilation
vasoconstriction

Quick Test

1. Why is it important that body temperature does not become too high?
2. What is the 'set point' for human body temperature?
3. Give **two** responses of the body to an increase in body temperature.
4. Why do humans shiver?

The Nervous System and the Eye

You must be able to:

- Explain the role of the different parts of the nervous system in responding to a stimulus
- Describe the functions of the main areas of the brain
- Explain how the eye can adjust to different light intensities and focus on objects at different distances.

The Nervous System

- The nervous system enables humans to react to their surroundings and coordinate their behaviour.
- Information from receptors passes to the **central nervous system (CNS)** (the brain and spinal cord).
- The CNS coordinates the response of effectors, i.e. muscles contracting or glands secreting hormones.
- Reflex actions are automatic and rapid so they can protect the body. They do not involve the conscious part of the brain:
  1. The pain stimulus is detected by receptors.
  2. Impulses from the receptor pass along a sensory neurone to the CNS.
  3. An impulse then passes through a relay neurone.
  4. A motor neurone carries an impulse to the effector.
  5. The effector (usually a muscle) responds, e.g. to withdraw a limb away from the source of pain.
- Neurones are not directly connected to each other.
- They communicate with each other via **synapses** (gaps between neurones).

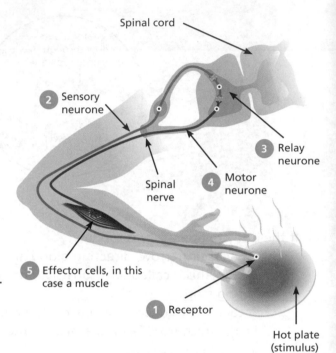

Spinal cord

2 Sensory neurone

3 Relay neurone

4 Motor neurone

Spinal nerve

5 Effector cells, in this case a muscle

1 Receptor

Hot plate (stimulus)

Key Point

When an electrical impulse reaches a synapse, a chemical is released that diffuses across the gap between the two neurones. This causes an electrical impulse to be generated in the second neurone.

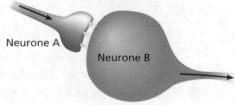

Neurone A

Neurone B

REQUIRED PRACTICAL	
Investigating the effect of a factor on human reaction time.	
Sample Method Reaction time can be investigated by seeing how quickly a dropped ruler can be caught between finger and thumb: 1. The experimenter holds a metre ruler from the end. 2. The subject has their finger and thumb a small distance apart, either side of the ruler, on the 50cm line. 3. The experimenter lets go of the ruler and the subject has to trap it. 4. The distance the ruler travels from the 50cm line is noted. 5. The experiment is repeated on subjects that have just drunk coffee or cola and subjects that have not.	**Considerations, Mistakes and Errors** - It is very difficult to control the variables in this experiment. - To obtain reliable results, large numbers of subjects need to be tested and averages taken.
Variables - The independent variable is whether the subject has taken in caffeine or not. - The dependent variable is the distance that the ruler travels. - The control variables are the age, sex and mass of the subjects.	**Hazards and risks** - There are limited risks with this experiment.

The Brain and the Eye

- The brain controls complex behaviour.
- It is made of billions of interconnected neurones and has different regions that carry out different functions.
- Three of these main regions are the: **cerebral cortex**, **cerebellum** and **medulla**.

HT Neuroscientists have been able to map the regions of the brain to particular functions by:
  - studying patients with brain damage
  - electrically stimulating different parts of the brain
  - using MRI scanning techniques.

HT The complexity and delicacy of the brain makes investigating and treating brain disorders very difficult.

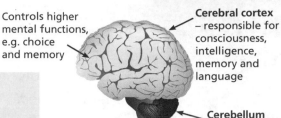

Controls higher mental functions, e.g. choice and memory

Cerebral cortex – responsible for consciousness, intelligence, memory and language

Cerebellum – coordinates movement and balance

Medulla – controls automatic actions such as heartbeat and breathing

- The eye is a sense organ. In it, the:
  - **retina** contains receptor cells that are sensitive to the brightness and colour of light
  - **optic nerve** carries impulses from the retina to the brain
  - **sclera** forms a tough outer layer, with a transparent region at the front called the **cornea**
  - **iris** controls the size of the **pupil** and the amount of light reaching the retina
  - **ciliary muscles** and **suspensory ligaments** can change the shape of the lens to focus light onto the retina.
- **Accommodation** is the process of changing the shape of the lens to focus on near or distant objects.

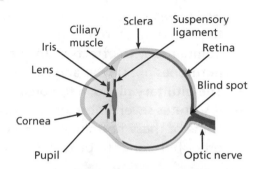

To focus on a near object the:
- ciliary muscles contract
- suspensory ligaments loosen
- lens becomes thicker and refracts light rays strongly.

Focus on a Near Object

Ciliary muscle

Suspensory ligament

Lens

Focus on a Distant Object

Ciliary muscle

Suspensory ligament

Lens

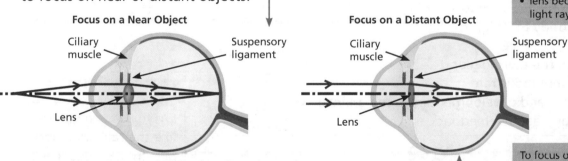

To focus on a distant object the:
- ciliary muscles relax
- suspensory ligaments are pulled tight
- lens is pulled thin and only slightly refracts light rays.

- Two common defects of the eyes occur when rays of light do not focus on the retina: **myopia** (short sightedness) and **hyperopia** (long sightedness). You may be required to interpret ray diagrams to explain how corrective lenses work.
- Generally these defects are treated with spectacle lenses, which refract the light rays so that they focus on the retina.
- New technologies include contact lenses, laser surgery to change the shape of the cornea, and replacement eye lenses.

Quick Test

1. Which type of neurone is responsible for sending impulses from the receptors to the CNS?
2. When your eyes are exposed to a bright light your pupils automatically become smaller. Why is this?
3. What is the gap between two neurones called?

Key Words

central nervous system (CNS)	iris
	pupil
synapse	ciliary muscles
cerebral cortex	suspensory
cerebellum	ligaments
medulla	accommodation
retina	refract
optic nerve	myopia
sclera	hyperopia
cornea	

Hormones and Homeostasis

You must be able to:

- Describe the principles of hormonal coordination and control
- Describe the location of the main hormone-producing glands
- Explain how hormones are used to control blood glucose levels
- Describe how urea is produced and excreted and how water balance is maintained.

The Endocrine System

- The **endocrine system** is made up of glands that secrete hormones directly into the bloodstream.
- **Hormones** are chemical messengers that are carried in the blood to a target organ where they produce an effect.
- Compared with effects of the nervous system, the effects of hormones are slower and act for longer.
- The **pituitary gland** in the brain is a 'master gland'.
- It secretes several hormones in response to body conditions.
- Some of these hormones act on other glands to stimulate other hormones to be released and bring about effects.

Main Glands that Produce Hormones in the Human Body

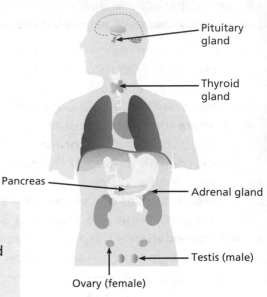

Pituitary gland

Thyroid gland

Pancreas

Adrenal gland

Testis (male)

Ovary (female)

HT **Adrenaline:**
- is produced by the adrenal glands in times of fear or stress
- increases the heart rate, boosting the delivery of oxygen and glucose to the brain and muscles
- prepares the body for 'flight or fight'.

HT **Thyroxine:**
- is produced by the thyroid gland
- increases the metabolic rate
- controls growth and development in young animals
- is controlled by negative feedback.

Control of Blood Glucose

- Blood glucose concentration is monitored and controlled by the pancreas.
- If the blood glucose concentration is too high:
  - the pancreas releases more of the hormone insulin
  - insulin causes glucose to move from the blood into the cells
  - in liver and muscle cells, excess glucose is converted to glycogen for storage.

HT If the blood glucose concentration is too low:
  - the pancreas releases glucagon
  - glucagon stimulates glycogen to be converted into glucose and released into the blood.

HT This is an example of negative feedback (see page 46).

- **Type 1 diabetes** is a disorder that:
  - is caused by the pancreas failing to produce sufficient insulin
  - results in uncontrolled high blood glucose levels
  - is normally treated with insulin injections.

Key Point

Enzymes from glands like the salivary gland pass into tubes called ducts. Endocrine glands are sometimes called ductless glands, because the hormones pass into the blood.

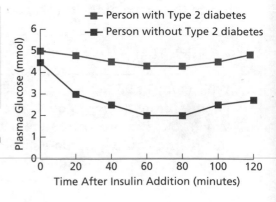

- Person with Type 2 diabetes
- Person without Type 2 diabetes

Time After Insulin Addition (minutes)

Plasma Glucose (mmol)

- **Type 2 diabetes** is a disorder that:
  - is caused by the body cells no longer responding to insulin
  - has obesity as a risk factor
  - is treated with a carbohydrate-controlled diet and regular exercise.

Water Balance

- Water leaves the body from the lungs during breathing, from the skin in sweat and in urine (along with ions and urea).
- If the concentration of the blood changes, then body cells will lose or gain too much water by osmosis.
- The balance of water and ions in the body is regulated by the kidneys.
- They also excrete **urea**, a waste product that is produced by the liver from the breakdown of proteins and contains nitrogen.

HT The digestion of proteins from food results in excess amino acids.

HT In the liver these excess amino acids are converted to ammonia in a process called **deamination**.

HT Ammonia is toxic, so it is immediately converted to urea and sent to the kidneys for safe excretion.

- The kidneys produce urine by:
  ① Filtering the blood.
  ② **Selective reabsorption** of useful substances, such as glucose, some ions and water.
  ③ This leaves urea and excess water and ions to form urine.
- These processes take place in millions of small tubes in the kidneys called **tubules**.

HT The water level in the body is controlled by the hormone **ADH**:
  - ADH is released by the pituitary gland when the blood is too concentrated.
  - It passes, in the blood, to the kidney tubules where it causes more water to be reabsorbed back into the blood.

- People who suffer from kidney failure may be treated by organ transplant or by using kidney **dialysis**.
- A dialysis machine takes over the role of the kidneys, it is used to remove waste products from the blood, three times a week.

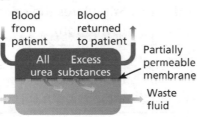

Blood from patient
Blood returned to patient
All urea
Excess substances
Partially permeable membrane
Dialysis fluid
Waste fluid

Kidney Tubules

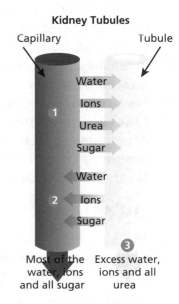

Capillary — Tubule

1
Water
Ions
Urea
Sugar

2
Water
Ions
Sugar

③

Most of the water, ions and all sugar — Excess water, ions and all urea

Key Words

endocrine system
hormone
pituitary gland
HT adrenaline
HT thyroxine
Type 1 diabetes
Type 2 diabetes
urea
HT deamination
selective reabsorption
tubules
HT ADH
dialysis

Quick Test

1. Where is the thyroid gland?
2. Where is insulin made?
3. What is urea?
4. HT If a person drinks a litre of water, what effect will this have on ADH release?

Hormones and Reproduction

You must be able to:

- Describe the roles of the main sex hormones in the body
- Explain how hormones control the menstrual cycle
- Describe the different methods of contraception
- HT Explain how hormones can be used to treat infertility.

The Sex Hormones

- Hormones play many roles in controlling human reproduction.
- During puberty, the sex hormones cause secondary sexual characteristics to develop.
- **Oestrogen**, from the ovaries, is the main female sex hormone.
- In females, at puberty, eggs begin to mature and be released. This is called **ovulation**.
- **Testosterone** is the main male sex hormone. It is produced by the testes and stimulates sperm production.
- After puberty, men produce sperm continuously, but women have a monthly cycle of events called the **menstrual cycle**.
- Several other hormones are involved in a woman's menstrual cycle.

Control of the Menstrual Cycle

- There are four hormones involved in control of the menstrual cycle:

Hormone	Secreted by	Function in the Menstrual Cycle
Follicle stimulating hormone (FSH)	Pituitary gland	• Causes eggs to mature in the ovaries in the first part of the cycle. HT Stimulates the ovaries to produce oestrogen.
Oestrogen	Ovaries	HT Inhibits FSH release. HT Stimulates LH release. • Makes the lining of the uterus grow again after menstruation.
Luteinising hormone (LH)	Pituitary gland	• Stimulates the release of the egg from the ovary (ovulation).
Progesterone	Empty follicle in the ovaries	• Maintains the lining of the uterus during the second half of the cycle. HT Inhibits both FSH and LH release.

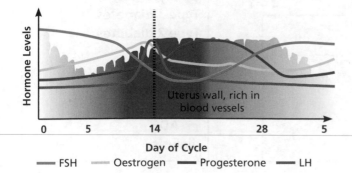

Hormone Levels

Uterus wall, rich in blood vessels

0   5   14   28   5

Day of Cycle

— FSH   — Oestrogen   — Progesterone   — LH

> ## Key Point
>
> An egg is released by ovulation approximately every 28 days, but this timescale can vary considerably between different women.

Reducing Fertility

- Fertility can be reduced by a variety of methods of **contraception**.
- Hormonal methods include:
    - oral contraceptives (the combined pill) that contain oestrogen and progesterone, which inhibit FSH production so that no eggs are released
    - an injection, implant or skin patch of slow release progesterone to stop the release of eggs for a number of months or years.
- Non-hormonal methods include:
    - barrier methods, such as condoms and diaphragms, that prevent the sperm from reaching an egg
    - intrauterine devices, which prevent embryos from implanting in the uterus
    - spermicidal creams, which kill or disable sperm
    - not having intercourse when an egg may be in the oviduct
    - surgical methods of male and female sterilisation, such as cutting the sperm ducts or tying the fallopian tubes.

HT Increasing Fertility

- Doctors may give FSH and LH in a **fertility drug** to a woman if her own level of FSH is too low to stimulate eggs to mature.
- **In vitro fertilisation** (IVF) treatment involves:
    - giving a woman FSH and LH to stimulate the growth of many eggs
    - collecting the eggs from the woman
    - fertilising the eggs with sperm from the father in the laboratory
    - inserting one or two embryos into the woman's uterus (womb).
- Fertility treatment gives a woman the chance to have a baby. However:
    - it is emotionally and physically stressful
    - the success rates are not high
    - it can lead to multiple births, which are a risk to both the babies and the mother.

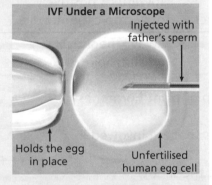

IVF Under a Microscope

Injected with father's sperm

Holds the egg in place

Unfertilised human egg cell

Key Point

Both contraception and fertility treatment are technological applications of science that raise ethical issues. People have many different views on these treatments.

Key Words

oestrogen
ovulation
testosterone
menstrual cycle
follicle stimulating hormone (FSH)
luteinising hormone (LH)
progesterone
contraception
HT fertility drug
HT in vitro fertilisation

Quick Test

1. What is the name of the male sex hormone?
2. What organ secretes FSH and LH?
3. Which hormone causes the release of an egg from the ovaries?
4. HT Where does fertilisation take place in IVF?

Plant Hormones

You must be able to:

- Describe the functions of plant hormones
- Explain how tropisms are controlled
- HT Describe how hormones can be used to manipulate plant growth.

Functions of Plant Hormones

- Plants can respond to changes in the external environment.
- The responses are usually slower than animal responses and include:
  - roots and shoots growing towards or away from a particular stimulus
  - plants flowering at a particular time
  - ripening of fruits.
- When part of a plant responds by growing in a particular direction, it is called a **tropism**.
- There are different types of tropism:

Stimulus	Growth of Shoots	Growth of Roots
Gravity	away = negatively **gravitropic** (geotropic)	towards = positively gravitropic (geotropic)
Light	towards = positively **phototropic**	away = negatively phototropic

- Tropisms are controlled by a group of plant hormones called **auxins**.
- Experiments have shown the steps that are involved in the response to light:
  1. More light reaches one side of the shoot.
  2. More auxin is sent down the shaded side of the shoot.
  3. This results in cells on the shaded side elongating more.
  4. The shoot, therefore, grows towards the light.
- In roots, auxin moves to the bottom of the root and causes less elongation, so the root grows downwards.

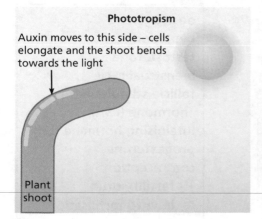

Phototropism

Auxin moves to this side – cells elongate and the shoot bends towards the light

Plant shoot

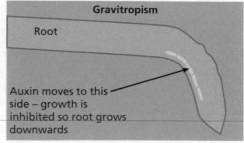

Gravitropism

Root

Auxin moves to this side – growth is inhibited so root grows downwards

HT Gibberellins are another group of plant hormones and are important in initiating seed germination.

HT Ethene is a gas and also a plant hormone. It controls cell division and ripening of fruits.

REQUIRED PRACTICAL	
Investigating the effect of light on the growth of newly germinated shoots.	
Sample Method 1. Take a number of small plastic boxes or containers and make a hole on one side. All the holes must be the same size. 2. Put some soil in the containers and plant 10 cress seeds in each. 3. Put a lid on each container. 4. Put the containers in even lighting, with the holes pointing in different directions. 5. Water the soil and leave for several days.	**Considerations, Mistakes and Errors** • It is important to make sure that the room is evenly lit with no windows. • The holes in the containers are positioned in different directions to show that only light entering is causing the shoots to bend
Variables • The independent variable is the position of the hole. • The dependent variable is the direction in which the seedlings grow. • The control variables are the number and type of seeds, soil type and volume of water.	**Hazards and risks** • There are limited risks with this experiment.

HT Uses of Plant Hormones

- Plant growth hormones are used by farmers and gardeners.
- Auxins are used:
  - as weed killers, because they make the weeds grow so rapidly that they use up their food reserves and die
  - as rooting powders, because they make cuttings produce roots when they are planted
  - for promoting growth in tissue culture (see page 83).
- Ethene is used in the food industry to control ripening of fruit during storage and transport.
- Gibberellins can be used to:
  - end seed dormancy and make seeds germinate
  - promote flowering, so that plants flower when there is the most demand, e.g. Easter and Mother's Day.
  - increase fruit size.

Quick Test

1. a) What is the name of the group of plant hormones that regulates tropisms?
   b) What effect do these hormones have on plant cells in the shoot?
2. What name is given to a shoot's response to light from a particular direction?
3. **HT** What do selective weed killers do?
4. **HT** Ripe fruits give off ethene. Why are green tomatoes often kept with a red tomato?

 Key Point

 Key Point

Auxin weedkillers are selective. This means that they can be used on lawns to kill weeds but not the grass.

WEED AWAY
Kills weeds, not grass

 Key Point

Fruit can be picked unripe before it is transported long distances, so that it is not damaged. It can then be ripened using ethene when it reaches its destination.

Key Words

tropism
gravitropic
phototropic
auxins
HT gibberellins
HT ethene

Review Questions

Cell Structure

1. Use words from the box to complete the sentences.

cell wall	cytoplasm	in a nucleus	plasmids	flagella
free within the cell		guard cell	cytotoxin	

A bacterial cell consists of _____ surrounded by a cell membrane.

Outside the cell membrane is a _____.

The main chromosome in bacteria is found _____.

There are also small loops of DNA called _____. [4]

2. a) Where are proteins synthesised?
Tick **one** box.

In the chloroplasts ☐   In the mitochondria ☐

In the ribosomes ☐   In the vacuole ☐   [1]

b) Where do most of the chemical reactions in a cell take place?
Tick **one** box.

On the cell membrane ☐   In the vacuole ☐

In the cytoplasm ☐   In the ribosomes ☐   [1]

c) Where is energy from respiration released?
Tick **one** box.

Cytoplasm ☐   Mitochondria ☐

Cell wall ☐   Nucleus ☐   [1]

Total Marks _____ / 7

Investigating Cells

1 A group of students set up an experiment to investigate diffusion.

They make a model cell as shown in **Figure 1**.

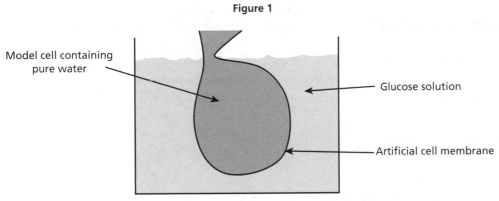

Figure 1

Model cell containing pure water

Glucose solution

Artificial cell membrane

The artificial membrane allows small molecules, such as water and glucose, to pass through.

The students fill the cell with pure water and place it in a solution containing glucose. They measure the concentration of glucose inside the cell every 20 minutes.

The graph in **Figure 2** shows the results.

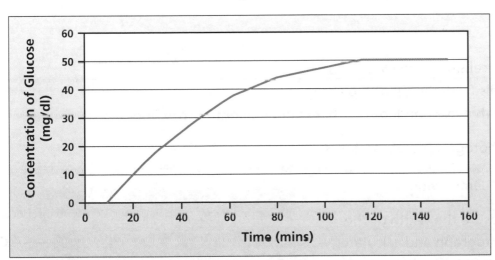

Figure 2

a) What was the concentration of glucose inside the cell after 40 minutes? [1]

b) Describe the pattern shown in the graph. [2]

c) Explain what happens to the model during the experiment. [3]

Review Questions

2 Fiona and Isaac are studying the diagram of a single-celled organism, called *Euglena*, shown in **Figure 3**.

a) Fiona finds out that *Euglena* is 0.02mm wide.

How wide is this in micrometres (μm)? [1]

b) Fiona thinks *Euglena* is a plant cell.

Why might she think this? [1]

c) Isaac says that *Euglena* cannot be a plant cell.

Suggest **one** reason why he might say this. [1]

d) *Euglena* reproduces by splitting into two in a process similar to mitosis.

What must happen in the nucleus of *Euglena* before it divides? [1]

Figure 3

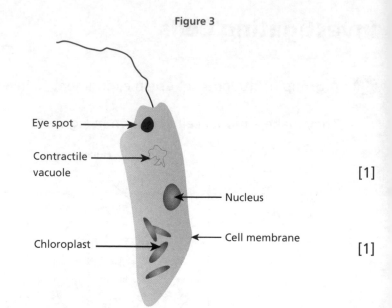

Total Marks _____ / 10

Cell Division

1 Karen wants to study cells dividing by mitosis.
She decides to use cells from the tip of a plant root.
She stains the cells with a dye and looks at them under a light microscope.

Figure 1 shows the photograph that she takes.

a) Cell **C** is actually 0.03mm wide.

Use a ruler to measure the width of the cell in Karen's photograph and calculate the magnification of his photograph.

$$\text{magnification} = \frac{\text{size of image}}{\text{size of real object}}$$
[2]

b) Which cell, **A**, **B** or **C** is about to divide?

You must explain your answer. [2]

Figure 1

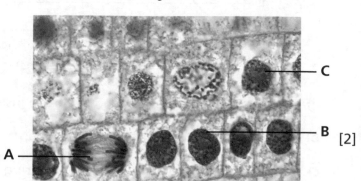

c) The cells in the photograph have not yet become specialised.

What name is given to cells that have not yet become specialised?
Tick **one** box.

Cloned ☐　　　　　Prokaryotic ☐

Mutated ☐　　　　　Undifferentiated ☐ [1]

d) i) Karen cannot see any ribosomes or mitochondria in these cells.

Explain why this is. [2]

ii) How could Karen take a photograph that would let her see these structures? [1]

Total Marks _____ / 8

Transport In and Out of Cells

1 A student carries out an investigation into the effect of different concentrations of sugar solution on the mass of potato chips.
The student weighs five chips separately and then places each one into a different concentration of sugar solution.
After one hour he removes the chips and then reweighs them.
Table 1 shows the results.

Table 1

Concentration of Sugar Solution (mol/dm³)	Mass of Chip at Start (g)	Mass of Chip After 1 hour (g)	Change in Mass (g)	Percentage Change in Mass
0.0	2.6	2.8	0.2	7.7
0.2	2.5	2.5	0.0	0.0
0.4	2.6	2.2	−0.4	−15.4
0.6	2.7	2.1	−0.6	

a) Work out the percentage change in mass for the potato chip in 0.6mol/dm³ sugar solution. [2]

b) Explain the result of the potato chip in 0.2mol/dm³ sugar solution. [3]

c) Before the student reweighs each chip he rolls it on some tissue paper.

Explain why he does this. [2]

Total Marks _____ / 7

Review Questions

Levels of Organisation

1 Organs are made up of a number of tissues.

Draw **one** line to join each type of tissue to its function.

Tissue	Function
glandular	can carry electrical impulses
nervous	can produce enzymes and hormones
muscular	a lining / covering tissue
epithelial	can contract to bring about movement

[3]

Total Marks _____ / 3

Digestion

1 **Figure 1** shows the human digestive system.

a) i) Which letter labels the organ that produces bile? [1]

ii) Which letter labels the place where digested food is absorbed into the bloodstream? [1]

iii) Which letter labels where protein digestion starts? [1]

b) The pancreas is a gland. It produces the hormone insulin.

Name **one** other substance produced by the pancreas. [1]

Figure 1

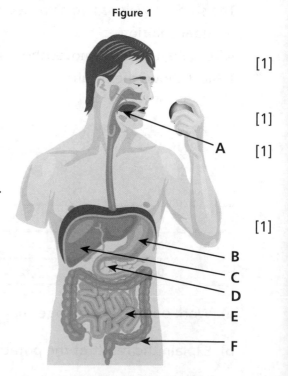

2 Different enzymes act on specific nutrients.

Draw **one** line to match each enzyme to the nutrient that it works on and draw **one** line to match each nutrient with its smaller subunit.

Enzyme	Nutrient	Subunit
protease	fats	glycerol and fatty acids
amylase	proteins	amino acids
lipase	starch	maltose [3]

Total Marks _____ / 7

Blood and the Circulation

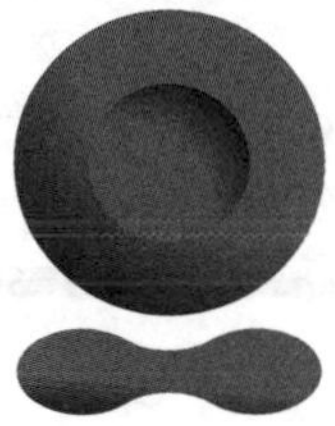

1 **Figure 1** shows a specialised cell.

Figure 1

a) What is the name of this cell? [1]

b) Why does this cell not have a nucleus? [1]

2 Which of the following are features of the alveoli?

Tick **two** boxes.

Good blood supply	☐	Poor blood supply	☐
No walls	☐	Thick walls	☐
Large surface area	☐	Small surface area	☐ [2]

3 a) Circle the correct options to complete the following sentences.

   i) The human circulation system is a **single / double / triple** system. [1]

   ii) Blood passes through the heart **once / twice / three times / four times** on each circuit. [1]

   iii) The blood vessels that carry blood away from the heart are called **arteries / veins / capillaries**. [1]

   b) Complete the following sentences.

   Plasma transports carbon dioxide from the _____ to the

   _____ and it transports _____ from the

   small intestine to the organs. [3]

4 Use words from the box to complete the sentences.

| an artery | an atrium | a valve | a vein |

a) A ventricle is filled with blood by the contraction of ....   [1]

b) When a ventricle contracts, blood is forced into ....   [1]

c) When a ventricle relaxes, blood is prevented from flowing back into it by the closing

of ....   [1]

Total Marks .... / 13

Non-Communicable Diseases

1 Statins are drugs prescribed to lower cholesterol levels in the blood.

a) What effect may high cholesterol levels have on blood vessels?   [2]

b) Describe why these changes to blood vessels can be dangerous.   [4]

Total Marks .... / 6

Transport in Plants

1 A group of students investigates the number of stomata on three different species of plants: **X**, **Y** and **Z**.

They estimate the number of stomata per cm² on both the upper and lower surfaces of the leaves.

Table 1 shows the results.

Table 1

Plant Species	Estimated Number of Stomata Per cm² of Leaf Surface	
	Upper Surface	Lower Surface
X	0	650
Y	3500	10 000
Z	6000	18 000

a) **Figure 1** shows a view of 0.001cm² of the lower surface of a leaf.

   i)   Calculate the number of stomata per cm² on this leaf. [2]

   ii) Based on your answer to part **i)**, which species (**X, Y** or **Z**) is this leaf likely to be from?

       Figure 1

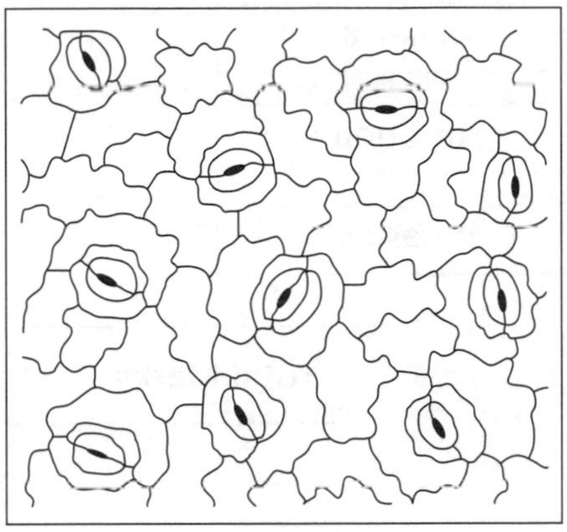

[1]

b) Which plant species (**X, Y** or **Z**) is likely to grow in a dry region?
You must explain your answer. [3]

c) Plants **X, Y** and **Z** all have more stomata on the lower surface of their leaves than the upper surface.

Suggest how this could help the plants to survive better. [2]

2 Use words from the box to complete the sentences.

gaps	epidermis	chloroplasts	phloem	water	mesophyll	gases

Like animals, plants have different tissues.

The tissue found covering the outer layers of the leaf is called the ....

The .... tissue carries out photosynthesis.

The cells of this tissue have .... between them to allow easy passage
of gases. [3]

3 Name the process by which water is lost from a leaf.
Draw a circle around your answer.

   osmosis   transpiration   circulation   diffusion   [1]

Total Marks .... / 12

Pathogens and Disease

1   Draw **one** line from each type of pathogen to the disease that it causes.

bacterium		rose black spot
virus		malaria
protist		salmonella
fungus		measles

[3]

> **Total Marks** .... / 3

Human Defences Against Disease

1   Children are vaccinated against a range of diseases.
Complete the sentences about vaccination.

Vaccines contain _____ or _____ pathogens.

These stimulate the white blood cells to produce _____ .

This results in the children becoming _____ to the disease.   [4]

2   The MMR vaccine was introduced in the UK in 1995.
It protects children from three diseases: measles, mumps and rubella.

a) What are the usual symptoms of measles?   [2]

b) Why is it important to vaccinate against measles?   [1]

c) The graph in **Figure 1** shows the uptake of the MMR vaccine in the UK between 1995
and 2007.

   i)   What percentage of 2-year-olds was vaccinated in 2007?   [1]

   ii) In which **two** years was there a significant decline in the uptake of the vaccine
   compared with the previous year?   [2]

   iii) Suggest a reason for this decline.   [1]

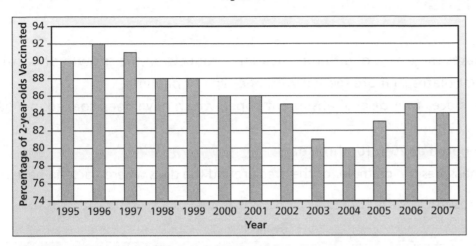

Figure 1

3 This is an extract from a leaflet about keeping healthy when travelling abroad.

> **In tropical areas, use an insect repellent.**
> **Check that you have been vaccinated against the diseases that you might come into**
> **contact with on your travels.**

a) Explain how using an insect repellent may help travellers to stay healthy. [3]

b) Explain why a person should be vaccinated before they go on holiday rather than
waiting to see if they get the disease before getting vaccinated. [3]

4 People have different opinions about vaccinations.
Some say it is better to give children a combined vaccination, i.e. one injection containing
several vaccines.
Others say that children should have several injections each containing a single vaccine.

Four people were asked their opinions on the subject.

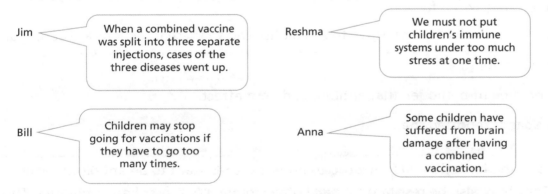

Jim: When a combined vaccine was split into three separate injections, cases of the three diseases went up.

Reshma: We must not put children's immune systems under too much stress at one time.

Bill: Children may stop going for vaccinations if they have to go too many times.

Anna: Some children have suffered from brain damage after having a combined vaccination.

Which **two** people think that **combined** vaccinations are a **good** idea? [2]

Total Marks _____ / 19

Treating Diseases

1 A student investigates the effect of different cleaning products on the growth of bacteria.

She sets up five agar plates that are seeded with one type of bacteria.

On each plate, she places four discs of absorbent paper, which have been soaked in a particular cleaning product.

The plates are incubated at 25°C for three days.

The student then measures the diameter of the area around the discs where bacteria have not grown.

Table 1 shows the results.

Table 1

Plate and Cleaning Product	Diameter of Area Around Disc Where Bacteria Did Not Grow (mm)				
	Disc 1	Disc 2	Disc 3	Disc 4	Mean
Plate 1 (soap)	1	2	2	1	1.5
Plate 2 (hand wash)	3	2	2	4	2.8
Plate 3 (kitchen cleaner)	5	4	6	5	5.0
Plate 4 (bathroom cleaner)	5	5	6	7	
Plate 5 (no cleaner)	0	0	0	0	0.0

a) Calculate the mean diameter of the area where bacteria did not grow for **Plate 4**. [2]
Give your answer to one decimal place.

b) Which product was the most effective at killing the bacteria? [1]

c) Why did the student use no cleaning product on **Plate 5**? [1]

d) What has the student done to make her results reliable? [1]

2 *Staphylococcus aureus* is a bacterium commonly found on human skin.

It can sometimes cause skin infections.

Most strains of this bacterium are sensitive to antibiotics and so these infections are easily cured.

a) i) What is an antibiotic? [2]

ii) Name one type of pathogen that antibiotics do not affect. [1]

b) A local newspaper reported that:

> 'Some strains of the bacterium *Staphylococcus aureus* are resistant to an antibiotic called methicillin. They may also be resistant to many other commonly prescribed antibiotics. This is called multiple resistance.
>
> It is difficult to get rid of these 'superbugs' when patients are infected with them.
>
> To try and reduce the risk of 'superbugs' appearing, we have to take care when using antibiotics.

i) What is the abbreviation used for multiple resistant strains of *Staphylococcus aureus* that are resistant to many antibiotics including methicillin? [1]

ii) List the precautions that are being taken with antibiotics to try and stop resistant strains developing. [3]

3 New drugs undergo testing before they are made available to the public.

a) The statements below describe the main stages in drug development and testing. They are in the wrong order.

Number the stages **1** to **5** to show the correct order. Stage 1 has been done for you.

The drug is passed for use on the general public.	
Trials using low doses of the drug take place on a small number of healthy volunteers.	
A new drug is made in the laboratory.	1
Clinical trials take place involving large numbers of patients and volunteers.	
The drug is tested in the laboratory using tissue culture.	

[3]

b) Suggest **two** reasons why it is necessary for new drugs to undergo such testing. [2]

c) In clinical trials, one group of patients is often given a placebo.

i) What is a placebo? [1]

ii) Explain why a placebo is given. [2]

Total Marks .... / 20

Plant Disease

1 **Figure 1** shows an aphid feeding on a plant stem.

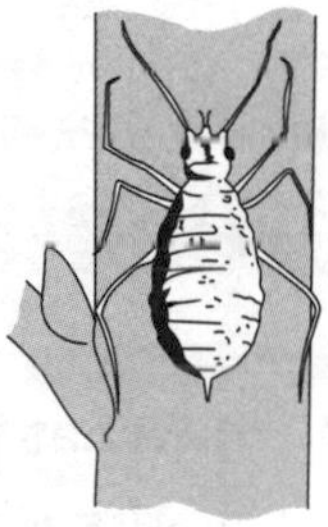

a) The aphid has mouthparts that allow it to feed from a tissue inside the stem.

Give the name of this tissue. [1]

Figure 1

b) Explain why many aphids feeding on a plant could stunt the growth of the plant. [2]

c) Explain how aphids could transmit pathogens from one plant to another. [2]

d) Before chemical pesticides became available, gardeners used to spray their plants with a solution made from tobacco plants.

Explain why they did this. [2]

Total Marks .... / 7

Practice Questions

Photosynthesis

1 **a)** Complete the **word** equation for photosynthesis.

carbon dioxide + _____ $\longrightarrow$ glucose + _____ [2]

b) Apart from the chemicals given in the equation above, what **two** other resources are required for photosynthesis? [2]

c) Plants need nitrates to produce proteins.

Where do plants obtain nitrates from?
Tick **one** box.

From the soil ☐   From oxygen ☐

From the leaves ☐   From photosynthesis ☐   [1]

2 A group of students carried out an experiment to prove that light is needed for photosynthesis. They selected one plant from their classroom window and covered a single leaf in foil. After several days the students tested the covered leaf and one uncovered leaf for starch using iodine solution.

a) Before testing the leaf for starch, the leaf has to be boiled in ethanol to remove the chlorophyll.

Suggest **two** safety precautions the students should take. [2]

b) What result would you expect from the leaf covered in foil? [1]

c) Explain why the students also tested an uncovered leaf. [2]

d) One student suggested that they should make their results more reliable.

How could they do this? [1]

Total Marks _____ / 11

Respiration and Exercise

1 **a)** What is the waste product of anaerobic respiration in animals?
Tick **one** box.

Carbon dioxide ☐   Lactic acid ☐

Hydrochloric acid ☐   Sulfuric acid ☐   [1]

b) List **three** differences between aerobic and anaerobic respiration. [3]

2 Circle the correct words to complete the paragraph.

When you exercise, your heart rate **decreases / increases**, which causes the flow of blood to your muscles to **decrease / increase**.

When you exercise, your breathing rate also **increases / decreases** to speed up the removal of **carbon dioxide / oxygen** from your muscles and the transport of **carbon dioxide / oxygen** to your muscles. [5]

3 Two students want to find out who is the fittest.
They carry out a simple investigation by performing star jumps for three minutes.
They record their pulse rate before the activity and every minute afterwards.

Table 1 shows the results.

Table 1

Time (mins)	Pulse Rate (bpm)	
	Student A	Student B
Before activity	68	72
1 minute after	116	160
2 minutes after	120	175
3 minutes after	116	168
4 minutes after	72	148
5 minutes after	66	92
6 minutes after	68	76

The results for Student A are plotted on the graph in **Figure 1** below.

a) Add the data for Student B to the graph. [2]

b) Suggest which student is fitter. You must give a reason for your answer. [2]

c) Both students experienced fatigue in their muscles.

Explain why this happens. [3]

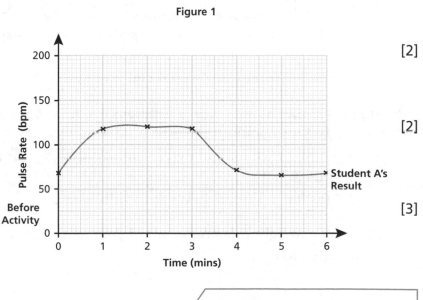

Figure 1

Total Marks _____ / 16

Practice Questions

Homeostasis and Body Temperature

1 A student in a school football team trained on a hot day and on a cold day.
On each day the student did the same amount of exercise and drank the same amount of water.

a) Circle the correct option to complete each sentence.

   i) On a hot day, the student would produce **less / more / the same amount of** urine. [1]

   ii) This is because they would produce **less / more / the same amount of** sweat. [1]

b) **HT** The student found that if they drank a greater volume of water it made them pass more urine.

   Explain why this is.
   Use ideas about blood concentration and ADH in your answer. [4]

Total Marks .... / 6

The Nervous System and the Eye

1 The nervous system allows organisms to react to their surroundings.

a) Put the following words in the correct order to show the pathway for receiving and responding to information. [3]

   | relay neurone | response | stimulus | receptor | sensory neurone |
   | effector | motor neurone |

b) Where in the pathway would you find a synapse? [2]

2 **Figure 1** shows how a myopia sufferer's eye fails to refract light on the retina properly.

Figure 1

a) On **Figure 1**, draw in the type of lens that could be used to solve this problem. Draw it in the correct position. [2]

b) Draw two light rays on the left of your lens to show how the rays are affected. [1]

3 **Figure 2** shows the side view of the human brain.

Figure 2

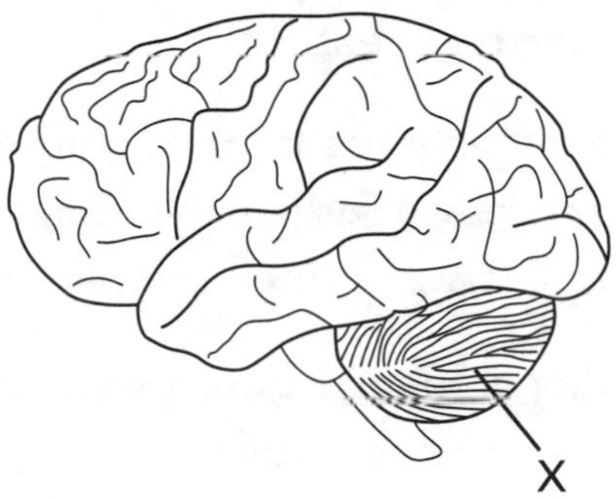

a) This person has a tumour and the area of the brain marked as **X** has been damaged.

What is the most likely effect of this damage on the person? [1]
Tick **one** box.

Loss of speech ☐   Loss of balance ☐

Loss of intelligence ☐   Loss of memory ☐

X

b) HT Explain why it is difficult for doctors to operate and remove tumours in the brain.   [2]

Total Marks _____ / 11

Hormones and Homeostasis

1 What is meant by the term 'selective reabsorption'?
Tick **one** box.

Excess substances are released. ☐

Water and small molecules are squeezed out of the blood. ☐

Useful substances are returned to the blood. ☐

The kidneys stop working. ☐   [1]

2 HT Urea must be removed from the body.

a) Which organ makes urea?   [1]

b) Which organ removes urea from the body?   [1]

c) What substance is broken down to form urea?   [1]

3 a) Complete the sentences.

In a dialysis machine, the blood flows through a _____ membrane.

Dialysis fluid contains the same concentrations of useful substances as _____,

so _____ and essential ions are not lost through diffusion.   [3]

b) Why does dialysis have to be repeated on a regular basis?

c) Give **two** precautions that are taken to prevent transplanted kidneys from being reject

4 A doctor gave a patient some glucose to eat.

He then measured the glucose and insulin levels in the patient's blood over a one-hour period.

The graph in **Figure 1** shows the doctor's results.

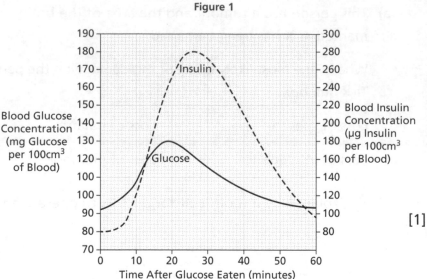

Figure 1

a) What was the level of glucose in the patient's blood before they ate the glucose? [1]

b) What was the maximum insulin level in the patient's blood during the one-hour period? [1]

c) When did the maximum level of insulin occur? [1]

d) Explain how the insulin in the patient's blood allows the glucose level to return to normal. [3]

e) The doctor performed this investigation to show that the patient did not have Type 1 diabetes.

Describe how the graph would differ if the patient did have Type 1 diabetes. [2]

Total Marks _____ / 18

Hormones and Reproduction

1 The first contraceptive pills contained large amounts of oestrogen.

a) Where in the body is oestrogen produced? [1]

b) Nowadays, contraceptive pills contain a much lower dose of oestrogen, which is combined with progesterone. Some birth control pills contain progesterone only.

Suggest **one** reason for the reduction of oestrogen levels in birth control pills. [1]

c) HT Which of the sentences about oestrogen is correct?
Tick **one** box.

Oestrogen speeds up the release of eggs. ☐

Oestrogen is an enzyme. ☐

Oestrogen inhibits the production of FSH. ☐ [1]

Total Marks _____ / 3

Plant Hormones

1 Ted uses ROOT-IT to promote root growth in his plant cuttings.
Bob makes his own rooting compound.

This is the method Bob uses:

1. Cut some twigs of willow tree and mash them into small pieces
2. Put the pieces of willow in a tub of water to stand overnight.
3. The next morning, remove the twigs and use the water as rooting compound.

Ted and Bob want to know whose rooting compound is best.
They take 10 geranium cuttings each, dip them in rooting compound and plant them.
They measure the height of each cutting every four days.

The graph in **Figure 1** shows the results.

Figure 1

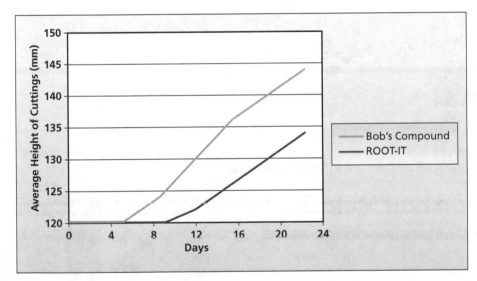

a) Give **two** variables that Ted and Bob need to control to make their experiment valid. [2]

b) What conclusion could you draw from the results of Bob and Ted's experiment? [2]

c) Suggest **one** disadvantage of using Bob's compound compared to ROOT-IT. [1]

d) What is the substance present in both compounds that promotes root growth? [1]

Total Marks .... / 6

Sexual and Asexual Reproduction

You must be able to:

- Describe some examples of asexual reproduction in different organisms
- Explain why sexual reproduction involves meiosis
- Explain why organisms may reproduce sexually or asexually at different times.

Asexual Reproduction

- **Asexual reproduction** involves:
  - only one parent
  - no fusion of **gametes**, so no mixing of genetic information
  - the production of genetically identical offspring (clones)
  - mitosis (see pages 12–13).
- Many plants reproduce asexually and in different ways, e.g.
  - strawberry plants send out long shoots called **runners**, which touch the ground and grow a new plant
  - daffodils produce lots of smaller bulbs, which can grow into new plants.

Daffodils

Bulb

Strawberry Plant

Runner

- Many fungi reproduce asexually by spores.
- Malarial protists reproduce asexually when they are in the human host.

Sexual Reproduction and Meiosis

- Sexual reproduction involves the fusion (joining) of male and female gametes:
  - sperm and egg cells in animals
  - pollen and egg cells in flowering plants.
- This leads to a mix of genetic information, which produces variation in the offspring.
- The formation of gametes involves **meiosis**.

Meiosis

| Cell with two pairs of chromosomes (diploid cell). | Each chromosome replicates itself. | Chromosomes part company and move to opposite poles. | Cell divides for the first time. | Copies now separate and the second cell division takes place. | Four haploid cells (gametes), each with half the number of chromosomes of the parent cell. |

- When a cell divides by meiosis:
  - copies of the genetic information are made
  - the cell divides twice to form four gametes, each with a single set of chromosomes
  - all gametes are genetically different from each other.
- Meiosis is important because it halves the number of chromosomes in gametes.
- This means that fertilisation can restore the full number of chromosomes.
- Once fertilised, the resulting cell divides rapidly by mitosis and cells become specialised by differentiation.

Sperm		Egg		Fertilised Egg Cell
23 chromosomes	+	23 chromosomes	=	46 chromosomes (23 pairs) – half from mother (egg) and half from father (sperm)

Asexual Versus Sexual Reproduction

- The advantages of sexual reproduction are:
  - it produces variation in the offspring
  - if the environment changes, any variation means that at least some organisms will be suited and can survive
  - it allows humans to selectively breed plants and animals, and increase food production (see pages 82–83).
- The advantages of asexual reproduction are:
  - only one parent is needed
  - it is more time and energy efficient, as the organism does not need to find a mate
  - it is faster than sexual reproduction
  - many identical offspring can be produced to make the best use of good conditions.
- Some organisms can reproduce both sexually and asexually:
  - many plants can produce seeds by sexual reproduction and can also reproduce asexually, e.g. using bulbs and runners
  - many fungi can make spores by asexual or by sexual reproduction
  - malaria parasites reproduce sexually in the mosquito as well as asexually in humans.

> **Key Point**
>
> If organisms have a choice, they often reproduce asexually when conditions are good to make lots of well-suited offspring. Sexual reproduction is used when conditions are getting worse, e.g. there is a drop in temperature or lack of food.

> **Quick Test**
>
> 1. What are strawberry runners?
> 2. How many gametes are made when one cell divides by meiosis?
> 3. The body cells of chickens have 78 chromosomes. How many chromosomes are in each gamete?
> 4. Which type of reproduction produces most variation?

> **Key Words**
>
> asexual reproduction
> gamete
> runners
> meiosis

DNA and Protein Synthesis

You must be able to:

- Describe how the genetic material is arranged in a cell
- Describe the structure of DNA
- **HT** Explain how DNA can code for proteins and how this can go wrong.

The Genome

- The genetic material in the nucleus of a cell is made of a chemical called **DNA**.
- The DNA is contained in structures called **chromosomes**.
- A **gene** is a small section of DNA on a chromosome.

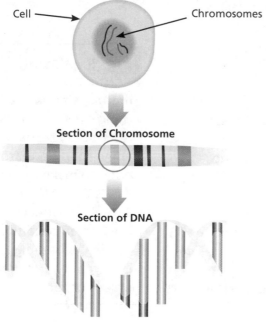

Cell

Chromosomes

Section of Chromosome

Section of DNA

- Each gene codes for a particular sequence of amino acids, to make a specific protein.
- The **genome** of an organism is its entire genetic material.
- The whole human genome has now been studied and this may have some important uses in the future, e.g.
  - doctors can search for genes linked to different types of disorder
  - it can help scientists to understand the cause of inherited disorders and how to treat them
  - scientists can investigate how humans may have changed over time, and even how ancient populations may have migrated across the globe.

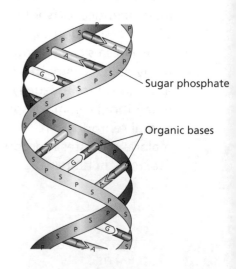

Sugar phosphate

Organic bases

A Single Nucleotide

Phosphate

Sugar

Organic base

The Structure of DNA

- DNA is a **polymer** made up of repeating units called **nucleotides**.
- Each nucleotide consists of:
  - a sugar
  - a phosphate
  - one of four bases: A, C, G or T.

- The nucleotides are joined together to form long strands.
- Each molecule has two alternating sugar and phosphate strands, which are twisted to form a double helix.
- Attached to each sugar is one of the four bases.

HT It is an attraction between the different bases that holds the two strands together:
- a C on one strand always links with a G on the opposite strand
- a T on one strand always links with an A on the opposite strand.

Making Proteins

- The order of bases on DNA controls the order in which amino acids are joined together to make a particular protein.
- A sequence of three bases is the code for one amino acid.

HT Proteins are synthesised on ribosomes using a template that has been taken from the DNA and carried out of the nucleus.
HT Carrier molecules then bring specific amino acids to add to the growing protein chain in the correct order.
HT When the protein chain is finished, it folds up to form a unique shape.
HT This unique shape allows the proteins to do their job as enzymes, hormones or structural proteins such as **collagen**.

HT **Mutations**

- A change in DNA structure is called a **mutation**.
- If any bases in the DNA are changed, then it may change the order of amino acids in the protein coded for by the gene.
- Mutations occur all the time. Most do not alter the protein, or only alter it slightly, so that it still works.
- A few mutations may cause the protein to have a different shape:
  - If it is an enzyme, then the substrate may no longer fit into the active site.
  - If it is a structural protein, it may lose its strength.
- Not all parts of DNA code for proteins:
  - Non-coding parts of DNA can switch genes on / off so that they can / cannot make specific proteins.
  - Mutations in these areas of DNA may affect how genes are expressed.

The order of bases in a section of DNA...

...controls the order in which amino acids...

...are joined together to form a protein.

> HT **Key Point**
>
> Certain chemicals and high-energy radiation can increase the rate at which mutations occur.

> HT **Key Point**
>
> Very occasionally, a mutation may occur that is useful. Without this type of variation, evolution by natural selection would not occur (see pages 80–81).

Key Words

DNA
chromosomes
gene
genome
polymer
nucleotide
HT collagen
HT mutation

Quick Test

1. What are chromosomes made from?
2. What is a section of chromosome called?
3. Name the **three** types of molecule that make up DNA.
4. HT What is a mutation?
5. HT Give a function of non-coding sections of DNA.

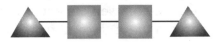

Patterns of Inheritance

You must be able to:

- Describe the contribution made by Gregor Mendel to the study of genetics
- Explain how ideas about genetics have changed since his work
- Predict the outcome of genetic crosses using genetic diagrams
- Describe examples of human genetic disorders
- Explain how sex is determined in humans.

Gregor Mendel

- Until the mid-19th century, most people thought that sexual reproduction produced a blend of characteristics, e.g. if a red flowering plant was crossed with a white flowering plant, then pink flowering plants were produced.
- Gregor Mendel investigated this by carrying out breeding experiments on pea plants.
- He found that characteristics are determined by 'units' that are inherited (passed on) and do **not** blend together.
- Later in the 19th century, the behaviour of chromosomes during cell division was observed.
- Then in the early 20th century, scientists realised that chromosomes and Mendel's 'units' behaved in similar ways.
- They decided that the 'units', now called genes, were located on chromosomes.
- In the mid-20th century, scientists worked out what the structure of DNA looked like and the mechanism by which genes work.
- The importance of Mendel's discovery was not recognised during his lifetime because:
  - he was a monk working in a monastery, not a scientist at a university
  - he did not publish his work in a well-known book or journal.

> ### Key Point
>
> The development of the gene theory, explaining how characteristics are passed on, is a good example of how ideas gradually change and develop. It illustrates how scientists make new observations and discoveries over time.

Modern Ideas About Genetics

- Some characteristics are controlled by a single gene, e.g. fur colour in mice and red-green colour blindness in humans.
- Each gene may have different forms called **alleles**, e.g. the gene for the attachment of earlobes has two alleles – attached or free.
- An individual always has two alleles for each gene:
  - One allele comes from the mother.
  - One allele comes from the father.
- The combination of alleles present in a gene is called the **genotype**, e.g. bb.
- How the alleles are expressed (what characteristic appears) is called the **phenotype**, e.g. blue eyes.
- Alleles can either be **dominant** or **recessive**.
- If the two alleles present are the same, the person is **homozygous** for that gene, e.g. BB or bb.
- If the alleles are different, they are **heterozygous**, e.g. Bb.

> ### Key Point
>
> A dominant allele is always expressed, even if only one copy is present. A recessive allele is only expressed if two copies are present, i.e. no dominant allele is present.

Genetic Crosses

- Most characteristics are controlled by several genes working together.
- If only one gene is involved, it is called **monohybrid inheritance**.
- Genetic diagrams or **Punnett squares** can be used to predict the outcome of a monohybrid cross.
- These diagrams use: capital letters for dominant alleles and lower case letters for recessive alleles.
- For example, for earlobes:
  - the allele for a free lobe is dominant, so **E** can be used
  - the allele for an unattached lobe is recessive, so **e** can be used.
- These Punnett squares show the possible outcomes of three crosses:

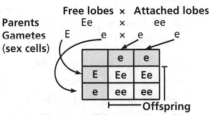

Each offspring will have a 1 in 2 chance of having attached lobes (because the dominant allele is present in half the crosses).

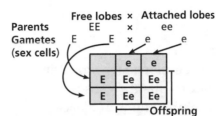

Each offspring will have free lobes (because the dominant allele is present in each cross).

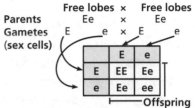

Each offspring has a 3 in 4 chance of having free lobes (because the dominant allele is present in three out of four crosses).

Genetic Disorders

- Some human disorders are inherited and are caused by the inheritance of certain alleles:
  - **Polydactyly** (having extra fingers or toes) is caused by a dominant allele.
  - **Cystic fibrosis** (a disorder of cell membranes) is caused by a recessive allele.

Sex Determination

- Only one pair out of the 23 pairs of chromosomes in the human body carries the genes that determine sex.
- These are called the **sex chromosomes**.
- In females, the two sex chromosomes are identical and are called X chromosomes (XX).
- Males inherit an X chromosome and a much shorter chromosome, called a Y chromosome (XY).
- As with all chromosomes, offspring inherit:
  - one sex chromosome from the mother (X)
  - one sex chromosome from the father (X or Y).

> ## Key Point
>
> It is now possible to test unborn foetuses for a range of genetic disorders. However, this means that the parents may have to make difficult decisions about the future of their baby. There is also a small risk to the pregnancy when removing foetal cells to test.

> ## Key Words
>
> allele
> genotype
> phenotype
> dominant
> recessive
> homozygous
> heterozygous
> monohybrid inheritance
> Punnett square
> polydactyly
> cystic fibrosis
> sex chromosomes

> ## Quick Test
>
> 1. What are the different forms of a gene called?
> 2. What do we call a combination of one dominant and one recessive allele?
> 3. Does a recessive or dominant allele cause cystic fibrosis?
> 4. What sex chromosomes are present in a male liver cell?

Variation and Evolution

You must be able to:

- Describe the main sources of variation between individuals
- Explain Darwin's theory of natural selection
- Describe some of the evidence for evolution.

Variation

- In a population, differences in the characteristics of individuals are called **variation**.
- This variation may be due to differences in:
  - the genes individuals have inherited (genetics)
  - the conditions in which individuals have developed (environment)
  - a combination of both genetic and environmental causes.
- Sexual reproduction produces different combinations of alleles and, therefore, variation, but only mutations create new alleles.
- Whereas most mutations do not affect the phenotype, a small number do. Within these, very rarely, a mutation may produce a phenotype that gives an organism a great survival advantage.

Natural Selection

- **Evolution** is the gradual change in the inherited characteristics of a population over time. This may lead to the formation of a new species.
- Many people have put forward theories to explain evolution.
- The theory that most scientists support is called **natural selection**, which was put forward by Charles Darwin.
- It states that all species have evolved from simple life forms that first developed more than three billion years ago.
- During a round-the-world expedition, Darwin observed:
  - Organisms often produce large numbers of offspring.
  - Populations usually stay about the same size.
  - Organisms are all slightly different – they show variation.
  - Characteristics can be inherited.
- Darwin used his observations to make these conclusions:
  - There is a struggle for existence.
  - More organisms are born than can survive.
  - The ones that survive and breed are the ones best-suited to the environment.
  - They pass on their characteristics to their offspring.
  - Over long periods of time the characteristics of populations change.
- In 1858, Alfred Russel Wallace suggested a similar theory to Darwin and this made Darwin realise that he should publish his ideas.
- Darwin published his theory in a book called *On the Origin of Species* in 1859.
- There was a lot of controversy over Darwin's ideas and several reasons why it took a long time for people to accept his theory:
  - The theory challenged the idea that God made all the organisms that live on Earth.

> ### Key Point
>
> ABO blood groups are controlled by a single gene, but height is the result of a combination of genes and environment. That is why there is a wide spread of possible heights.

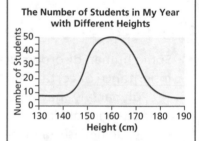

The Number of Students in My Year with Different Heights

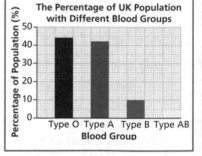

The Percentage of UK Population with Different Blood Groups

- There was not enough evidence at the time the theory was published to convince many scientists.
- The mechanism of inheritance and variation was not known until 50 years after the theory was published.
- There were other theories, including that of Jean-Baptiste Lamarck.
- Lamarck's theory was based on the idea that changes that occur in an organism during its lifetime can be inherited.
- We now know that, in most cases, this type of inheritance cannot occur.

Evidence for Evolution

- One problem with Darwin's theory has now been solved – we now know the mechanism of inheritance and variation, i.e. that characteristics are passed on to offspring in genes.
- There is also some evidence for evolution provided by **fossils**.
- Fossils are the remains of organisms from hundreds of thousands of years ago that are found in rocks.
- Fossils may be formed in various ways:
  - from the hard parts of animals that do not decay easily
  - from parts of organisms that have not decayed, because one or more of the conditions needed for decay was absent
  - when parts of the organisms are replaced by other materials as they decay
  - as preserved traces of organisms, e.g. footprints, burrows and root pathways.
- Scientists have used fossils to look at how organisms have gradually changed over long periods of time.
- Although fossils have been useful to scientists there are problems.
- There are gaps in the fossil record, because:
  - Many early forms of life were soft-bodied, which means that they have left very few traces behind.
  - What traces there were may have been destroyed by geological activity.
- The development of antibiotic-resistant strains of bacteria can be explained using the theory of natural selection:
  - Bacteria can evolve rapidly because they reproduce at a fast rate.
  - When they reproduce, mutations occur.
  - Some mutated bacteria might be resistant to antibiotics and are not killed.
  - These bacteria survive and reproduce, so a resistant strain develops (see pages 38–39).
- There is still much debate today among scientists over the theory of evolution and the origins of life.

Lamarck believed that the necks of giraffes stretched during their lifetime to reach food in trees. They then passed this characteristic on to the next generation.

Darwin believed giraffes that had longer necks could reach more food in trees, so they were more likely to survive and reproduce.

Key Point

When scientists describe natural selection now, they can talk about alleles being passed on, which will cause changes to the phenotypes in a population.

Quick Test

1. Who developed the theory of evolution by natural selection?
2. Give **one** reason why the theory of evolution by natural selection was not accepted by Darwin's peers straight away.
3. Who described a theory of evolution where changes during the lifetime of an organism were passed on to the next generation?
4. Give **one** reason why there are gaps in the fossil record.

Key Words

variation
evolution
natural selection
fossils

Manipulating Genes

You must be able to:

- Describe the process of selective breeding
- Explain how genetic engineering can be used to change organisms' characteristics
- Compare different cloning techniques.

Selective Breeding

- Humans have been using **selective breeding**, or artificial selection, for thousands of years to produce:
  - food crops from wild plants
  - domesticated animals from wild animals.
- It is the process by which humans breed plants and animals with particular, desirable genetic characteristics.
- Selective breeding involves several steps:
  1. Choose parents that best show the desired characteristic.
  2. Breed them together.
  3. From the offspring, again choose those with the desired characteristic and breed.
  4. Continue over many generations.
- The type of characteristic that could be selected includes:
  - disease resistance in food crops
  - animals that produce more meat or milk
  - domestic dogs with a gentle nature
  - large or unusual flowers.
- However, selective breeding can lead to 'inbreeding', where some breeds are particularly prone to disease or inherited defects.

Example of Selective Breeding

Choose the spottiest two to breed...

... and then the spottiest of their offspring...

... to eventually get Dalmatians.

Genetic Engineering

- **Genetic engineering** is a more recent way of bringing about changes in organisms.
- It involves changing the characteristics of an organism by introducing a gene from another organism.

HT In genetic engineering:
1. Enzymes are used to isolate the required gene.
2. This gene is inserted into a vector, e.g. a bacterial plasmid or virus.
3. The vector is used to insert the gene into the required cells.

HT If the genes are put into the cells of animals or plants at the egg or embryo stage, then all cells in the organism will get the new gene.

- Plant crops have been genetically engineered to:
  - be resistant to diseases, insects or herbicide attack
  - produce bigger, better fruits.
- Crops that have had their genes modified in this way are called **genetically modified (GM)** crops.

Key Point

Owners have to be very careful when mating pedigree dogs to make sure that they are not too closely related.

- Some people are concerned about GM crops and the possible long-term effects on populations of wild flowers and insects and on human health (if consumed).
- Other ethical considerations include the role that multinational companies play in manufacturing GM crops and setting the price.
- Fungi or bacterial cells have been genetically engineered to produce useful substances, e.g. human insulin to treat Type 1 diabetes.

Part of a human chromosome

Human insulin gene

Ring of bacterial DNA cut open

Human insulin gene inserted into bacterial DNA

- In the future, it may be possible to use genetic modification to cure or prevent some inherited diseases in humans.

Cloning

- **Clones** are genetically identical individuals.
- They are produced naturally by asexual reproduction (page 74–75).
- There are also a number of artificial ways of making clones.
- In plants, identical plants can be produced from:
    - **cuttings** – this is a method often used by gardeners
    - **tissue culture** – this uses small groups of cells to grow new plants and is used commercially and to preserve rare plant species.
- In animals, clones can be produced by splitting apart cells from an embryo before they become specialised, then transplanting the identical embryos into host mothers.
- It has also become possible to produce clones using adult cells:
    1 Remove the nucleus from an unfertilised egg cell.
    2 Insert the nucleus from an adult body cell of the organism you want to clone into the empty egg cell.
    3 Stimulate the egg cell to divide using an electric shock.
    4 Allow the resulting embryo to develop into a ball of cells.
    5 Insert the embryo into the womb of a **surrogate** female to continue its development.

1 Select a plant

2 Take cuttings

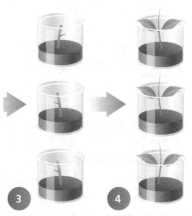

3 Place in damp atmosphere

4 New genetically identical plants develop

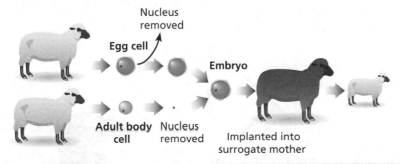

Nucleus removed

Egg cell

Embryo

Adult body cell

Nucleus removed

Implanted into surrogate mother

Quick Test

1. State **one** characteristic that farmers might selectively breed cows for?
2. HT What is used to cut genes from chromosomes in genetic engineering?
3. Describe **one** concern about GM crops.
4. What is an electric shock used for in cloning?

Key Words

selective breeding
genetic engineering
genetically modified (GM)
clone
cuttings
tissue culture
surrogate

Classification

You must be able to:

- Explain why classification systems have changed over time
- Describe why organisms may become extinct
- Explain how new species are formed
- Describe how evolutionary trees are constructed.

Principles of Classification

- Traditionally, living things have been classified into groups based on their structure and characteristics.
- One of the main systems used was developed by Carl Linnaeus.
- Linnaeus classified living things into:
  kingdom → phylum → class → order → family → genus → species
- Organisms are named by the **binomial system**, i.e. they have two parts to their Latin name:
  - The first part is their **genus**.
  - The second part is their **species**.
- New models of classification were proposed because:
  - microscopes improved, so scientists learnt more about cells
  - biochemical processes became better understood.
- Due to evidence, e.g. from genetic studies, there is now a **three-domain system** developed by Carl Woese.
- In this system organisms are divided into:
  - archaea (primitive bacteria, usually living in extreme environments)
  - bacteria (true bacteria)
  - eukaryota (including protists, fungi, plants and animals).

> ### Key Point
>
> The scientific name for lion is *Panthera leo* and for tiger it is *Panthera tigris*. This shows that they are in the same genus but different species. The cheetah is called *Acinonyx jubatus* – it is in a different genus and species.
>
>

Extinction

- Throughout the history of life on Earth, different organisms have been formed by evolution and some organisms have become **extinct**.
- Extinction may be caused by:
  - changes to the environment over geological (long periods of) time
  - new predators
  - new diseases
  - new, more successful competitors
  - a single catastrophic event, e.g. massive volcanic eruptions or collisions with asteroids.
- For example, the great auk is now extinct due to over-hunting.

Evolutionary Trees

- Evolutionary trees are a method used by scientists to show how they think organisms are related.
- They use current classification data for living organisms and fossil data for extinct organisms.

Great Auk

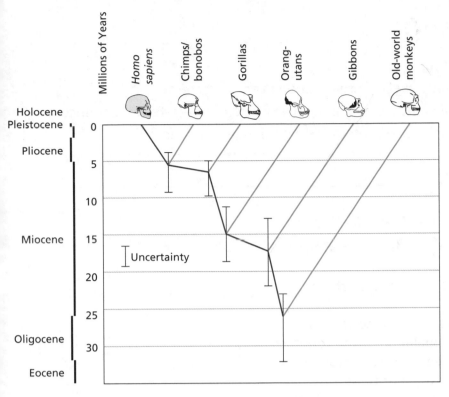

Speciation

- The lowest level in Linnaeus's classification system is the 'species'.
- Members of a species are similar enough to be able to breed with each other and produce fertile offspring.
- Alfred Russel Wallace, who proposed a theory for evolution, also worked on a theory of **speciation,** i.e. how new species develop:
  1. Populations become physically isolated from each other, e.g. by a mountain range or ocean.
  2. Genetic variation is present between the two populations.
  3. Natural selection operates differently in the two populations.
  4. The populations become so different that successful interbreeding is no longer possible.
- Wallace's theory is supported by recent studies.

Key Point

Darwin and Wallace both visited many islands before putting forward their theories. New species often form on islands, because they are isolated from populations on other islands.

Quick Test

1. What classification group comes between 'class' and 'family' in Linnaeus' system?
2. How many domains are there in the current classification system?
3. The binomial name for domestic cats is *Felis catus*. What does this tell you about how they are classified?
4. Give **two** reasons why a species may become extinct.
5. How can scientists tell if two groups or organisms have become different species?

Key Words

binomial system
genus
species
three-domain system
extinct
speciation

Ecosystems

You must be able to:

- Describe different levels of organisation in an ecosystem
- Explain why different organisms in an ecosystem show interdependence
- Describe the factors that determine where organisms can live
- Describe some of the techniques used by scientists to study ecosystems.

Relationships Between Organisms

- An **ecosystem** is all the organisms living in a habitat *and* the non-living parts of the habitat.
- There are different levels of organisation in an ecosystem:
  - individual organisms
  - populations – groups of individuals of the same species
  - communities – made up of many populations living together.
- To survive and reproduce, organisms require certain resources from their habitat and the other living organisms there.
- Trying to get enough of these resources results in **competition**.
- Plants in a community or habitat often compete with each other for light, water, space and mineral ions from the soil.
- Animals often compete with each other for food, mates and territory.
- As well as competing with each other, species also rely on each other for food, shelter, pollination, seed dispersal, etc. This is called **interdependence**.
- Because of interdependence, removing one species from a habitat can affect the whole community.
- In a stable community, all the species and environmental factors are in balance so that population sizes stay fairly constant.
- Tropical rainforests and ancient oak woodlands are examples of stable communities.

Adaptations

- Factors that can affect communities can be **abiotic** (non-living) or **biotic** (living).
- Abiotic factors include:
  - light intensity
  - temperature
  - moisture levels
  - soil pH and mineral content
  - wind intensity and direction
  - carbon dioxide levels for plants
  - oxygen levels for aquatic animals.
- Biotic factors include:
  - availability of food
  - new predators arriving
  - new pathogens / diseases
  - one species outcompeting another.

Key Point

All ecosystems should be self-supporting, but they do need energy. Energy is usually transferred into the ecosystem as light energy for photosynthesis in green plants, which are the start of the food chain.

- Organisms have **adaptations** (features) that enable them to survive in the conditions in which they normally live.
- These adaptations may be structural, behavioural or functional.
- Some organisms live in environments that are very extreme, e.g. with high temperature, pressure or salt concentration. These organisms are called **extremophiles**.
- Bacteria living in deep-sea vents are extremophiles.

Studying Ecosystems

- A group of organisms of one species living in a habitat is called a **population**.
- Scientists often want to estimate the size of a population.
- This might involve sampling using a square frame called a **quadrat**.

Cactus

Needles instead of leaves reduce water loss

Thick stem stores water

Extensive root system to take in water

REQUIRED PRACTICAL	
Investigating the population size of a common species in a habitat.	
Sample Method 1. Place a quadrat on the ground at random. 2. Count the number of individual plants of one species in the quadrat. 3. Repeat this process a number of times and work out the mean number of plants. 4. Work out the mean number of plants in 1m². 5. Measure the area of the whole habitat and multiply the number of plants in 1m² by the whole area.	**Considerations, Mistakes and Errors** • The main consideration in the experiment is making sure that the quadrats are placed at random. Using random numbers to act as coordinates can help with this. • The more samples that are taken, then the more accurate the estimate should be.
Variables • The dependent variable is the number of plants in the quadrat.	**Hazards and Risks** • Care should be taken to wash hands after ecology work in a habitat. • Care should be taken throwing quadrats – throw low to the ground, not up in the air.

- To see how plants are spread or distributed in a habitat:
  ① Stretch a long tape, called a **transect line**, across the area.
  ② Place a quadrat down at regular intervals along the line.
  ③ Count the plants in the quadrat each time.

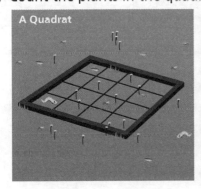

A Quadrat

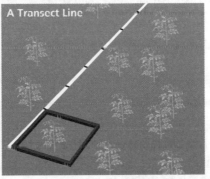

A Transect Line

Quick Test

1. Name **one** stable community.
2. What resources do plants compete for?
3. What are extremophiles?
4. What is a quadrat?
5. How can an estimate of a population be made more accurate?

Cycles and Feeding Relationships

You must be able to:

- Describe the factors needed for decomposition
- Explain how carbon and water are recycled in nature
- Explain how feeding relationships can be shown by food chains and predator–prey graphs.

Decomposition

- When organisms die or produce waste products, the dead material is broken down by organisms called **decomposers**.
- Decomposers are certain types of bacteria and fungi.
- Decomposers need oxygen, moisture, a suitable temperature and a suitable pH to break down waste.
- Decomposers break down dead waste by secreting enzymes, which partly digest the waste.
- The decomposers then take up the small, soluble food molecules.
- In a compost heap, gardeners try to provide optimum conditions for decay.
- The compost produced is used as a natural fertiliser for growing plants.
- If waste is broken down in anaerobic conditions, methane gas is produced.
- Biogas generators can be used to produce biogas from waste for use as a fuel.

Key Point

While gardeners want to encourage decay in compost heaps, the food industry wants to stop or slow down the decay of food. Food preservation often involves removing one of the factors needed for decay.

REQUIRED PRACTICAL
Investigate the effect of temperature on the rate of decay of fresh milk by measuring pH change.

Sample Method	Considerations, Mistakes and Errors
1. Take a sample of fresh milk. 2. Place in a beaker and cover (**not** airtight). 3. Keep warm or at room temperature. 4. Measure the pH of the milk at regular intervals over several days.	• The pH can be measured using universal indicator paper. • A pH probe linked to a data logger will give a continuous read out of pH.

Recycling Materials

- All materials in the living world need to be recycled so that they can be used again in future organisms.
- The **carbon cycle** describes how carbon is recycled in nature.
- It relies on decomposers to return carbon to the atmosphere as carbon dioxide through respiration.
- The **water cycle** describes how fresh water circulates between living organisms, rivers and the sea.

The Carbon Cycle

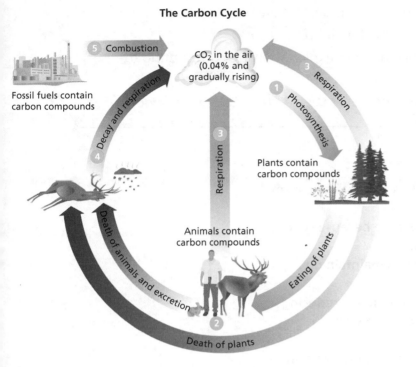

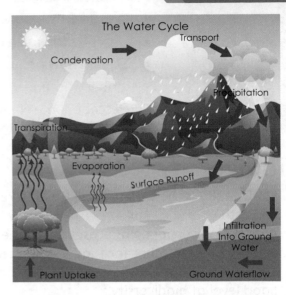

The Water Cycle

Feeding Relationships

- Feeding relationships in a community can be shown in **food chains**.
- All food chains begin with a **producer**, which synthesises (makes) molecules.
- The producer is usually a green plant, which makes glucose molecules by photosynthesis.
- Producers are eaten by primary consumers, which may be eaten by secondary consumers, which in turn may be eaten by tertiary consumers.
- Each of these feeding levels is called a **trophic level**.
- Trophic levels can be represented by numbers, starting at level one.
- Consumers that eat other animals are **predators** and those that are eaten are **prey**.
- Top consumers are **apex predators**. They are carnivores with no predators.
- In a stable community, the numbers of predators and prey rise and fall in cycles. This can be shown in a predator–prey graph.

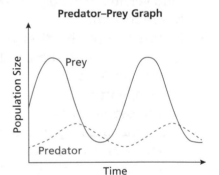

Predator–Prey Graph

Key Point

In the predator–prey graph, both lines follow the same pattern but the changes in predator numbers happen just after the changes in prey.

Quick Test

1. Why do gardeners often water their compost heaps in dry weather?
2. Why are crisp packets sometimes filled with nitrogen gas rather than air?
3. How can carbon from carbon dioxide in the air get back into compounds in living organisms?
4. What is a trophic level?
5. Why do predator numbers fall soon after a drop in prey numbers?

Key Words

decomposers
carbon cycle
water cycle
food chain
producer
trophic level
predator
prey
apex predator

Disrupting Ecosystems

You must be able to:

- Explain why biodiversity is so important and why it is at risk
- Describe the main causes of pollution
- Explain how pollution and over-exploitation are contributing to global warming
- Describe some of the steps that are being taken to maintain biodiversity.

Biodiversity

- **Biodiversity** is the variety of all the different species of organisms on Earth.
- A high biodiversity helps ecosystems to be stable because species depend on each other for food and shelter.
- The future of humans, as a species, relies on us maintaining a good level of biodiversity.
- Many human activities (see below) are responsible for reducing biodiversity, so action is now being taken to try to stop this reduction.
- Factors that put biodiversity at risk, and affect the distribution of species in an ecosystem, include changes in:
  - availability of water
  - temperature
  - atmospheric gases.
- These changes may be due to:
  - changes in the seasons
  - geographic activity, e.g. volcanoes or storms
  - human interaction.

Pollution

- **Pollution** kills plants and animals, which can reduce biodiversity.
- The human population is increasing rapidly and, in many areas, there is also an increase in the standard of living.
- This means that more resources are used and more waste is produced.
- Unless waste and chemical materials are properly handled, more pollution will be caused.
- Pollution can occur:
  - in water, from sewage, fertilisers or toxic chemicals
  - in air, from gases, e.g. sulfur dioxide, which dissolves in moisture in the atmosphere to produce **acid rain**
  - on land, from landfill and toxic chemicals, e.g. pesticides and herbicides, which may be washed from land into water.

Overexploitation

- Humans can also put biodiversity at risk by taking too many resources out of the environment.
- Building, quarrying, farming and dumping waste can all reduce the amount of land available for other animals and plants.
- Producing garden compost destroys peat bogs, reducing the area of this habitat and the variety of different organisms that live there.

> **Key Point**
>
> Many garden centres now sell 'peat-free' compost to try and reduce the destruction of peat bogs.

- The decay or burning of the peat releases carbon dioxide (a greenhouse gas) into the atmosphere.
- Cutting down trees and the destruction of forests is called **deforestation**.
- In tropical areas deforestation has occurred to:
  - provide land for cattle and rice fields to provide more food
  - grow crops from which biofuels can be produced.
- **Global warming** is a gradual increase in the temperature of the Earth.
- Many scientists think that it is being caused by changes in various gases, caused by pollution and deforestation.
- These gases include carbon dioxide and methane.
- There are a number of biological consequences of global warming:
  - loss of habitat, when low-lying areas are flooded by rising sea levels
  - changes in the distribution of species in areas where temperature or rainfall has changed
  - changes to the migration patterns of animals.

Conserving Biodiversity

- Scientists and governments have taken steps to reduce pollution and over-exploitation to help maintain biodiversity. These include:
  - setting up breeding programmes for endangered species
  - protecting rare habitats, e.g. coral reefs, mangroves and heathland
  - encouraging farmers to keep margins and hedgerows in fields
  - reducing deforestation and carbon dioxide emissions
  - recycling resources rather than dumping waste in landfill.

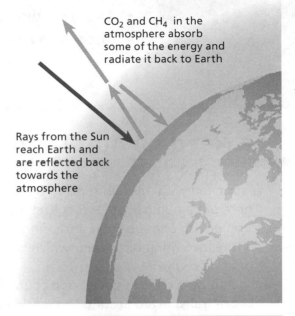

CO_2 and CH_4 in the atmosphere absorb some of the energy and radiate it back to Earth

Rays from the Sun reach Earth and are reflected back towards the atmosphere

GLASS PLASTIC PAPER

> **Key Point**
>
> Any increase in carbon dioxide levels will cause photosynthesis rates to increase. However, there is a limit to how well plants will be able to reduce the effect of global warming, especially if huge areas of rainforests are cut down.

Quick Test

1. Give **one** reason why the human population is producing more waste.
2. What can acidic pollutant gases cause?
3. Give **one** reason why deforestation is occurring.
4. What effect does deforestation have on the amount of carbon dioxide in the atmosphere and why?

> **Key Words**
>
> biodiversity
> pollution
> acid rain
> deforestation
> global warming

Feeding the World

You must be able to:

- Explain why providing enough food for everyone is becoming more difficult
- Explain how food production can be increased by manipulating energy flow
- Describe how biotechnology is used to increase food production.

The Need for More Food

- Food security involves making sure that all the world's population is supplied with enough food to be healthy.
- There are a number of factors that are making it harder to supply people with enough food, including:
  - the increasing birth rate, which has threatened food security in some countries
  - changing diets in developed countries, which mean some foods are transported around the world to meet demands
  - new pests and pathogens, which are affecting farming
  - changes in weather, which can affect food production, e.g. widespread famine due to drought
  - the cost of agricultural supplies such as fertilisers is increasing
  - conflicts in some parts of the world, which have affected supplies of water or food.
- Scientists and farmers are trying to find **sustainable** methods to feed all the people on Earth.
- Sustainable methods increase food production now, but leave enough resources for future generations.

Manipulating Energy Flow

- **Pyramids of biomass** can be used to compare the amount of biomass in each level of a food chain.
- They show the energy flow though ecosystems more clearly than food chains.
- Producers (green plants) transfer about 1% of incoming light energy via photosynthesis.
- They are in trophic level one, at the bottom of the pyramid.
- Pyramids of biomass are that particular shape because only about 10% of the biomass from each trophic level is transferred to the level above it.
- Losses of biomass are due to:
  - some of the food taken in being passed out of the body as faeces
  - large amounts of glucose are used in respiration
  - some material being lost in excretion, e.g. carbon dioxide and water in respiration, and water and urea in urine.
- This loss of biomass, and reduction in energy available, means that there are usually fewer organisms in the higher trophic levels.

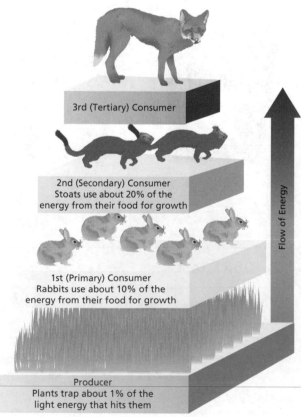

3rd (Tertiary) Consumer

2nd (Secondary) Consumer
Stoats use about 20% of the energy from their food for growth

1st (Primary) Consumer
Rabbits use about 10% of the energy from their food for growth

Producer
Plants trap about 1% of the light energy that hits them

Flow of Energy

- The efficiency of food production can be improved by reducing energy transfer from animals to the environment by:
  - limiting the movement of the animals
  - controlling the temperature of their surroundings.
- Factory farming uses these ideas, e.g. raising battery chickens in cages and calves in pens to limit their movement.
- Growth may also be promoted by feeding the animals high-protein foods.
- Fish are also grown in cages, which restrict their movement and allow farmers to feed them high-protein food.
- Fish stocks in the oceans are declining and some fish, such as cod in the northwest Atlantic, have reached very low levels.
- To try and stop fish numbers decreasing, regulations:
  - control the size of the nets used, so only older fish are caught
  - set fishing quotas, so that only a certain number can be caught.

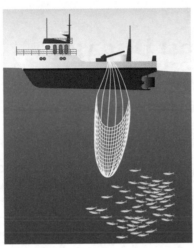

Biotechnology

- **Biotechnology** allows microorganisms to be grown in large quantities for food.
- They are grown in industrial-sized vats called **fermenters**, in which conditions are carefully controlled.
- The fungus *Fusarium* is useful for producing **mycoprotein**.
- Mycoprotein is:
  - protein-rich
  - low in fat
  - high in fibre
  - suitable for vegetarians.

- The fungus is grown on glucose syrup, in aerobic conditions, and then harvested and purified.
- Quorn™ is an example of a food containing mycoprotein.
- GM crops can be grown to provide more food or food with an improved nutritional value, e.g. golden rice.

Key Point

Golden rice is genetically engineered by inserting daffodil genes and bacterial genes into the rice genome to increase its vitamin A content. Hopefully, this will prevent large numbers of people from developing the vitamin A deficiency disease, night-blindness, which is common in Eastern Asia.

Quick Test

1. What is meant by the term 'sustainable' food production?
2. State **two** ways in which energy is transferred out of a food chain.
3. How are fishermen conserving stocks of fish?
4. What type of microorganism is used to make mycoprotein?

Key Words

sustainable
pyramid of biomass
biotechnology
fermenter
mycoprotein

Review Questions

Pathogens and Disease

1 **a)** Define the term 'vector'. [1]

b) What organism acts as a vector for the protist that causes malaria? [1]

c) Suggest **two** ways in which people can protect themselves from this vector. [2]

Total Marks _____ / 4

Human Defences Against Disease

1 **a)** Use words from the box to complete the sentences about vaccination.

Each word may be used once, more than once or not at all.

| antibodies | antibiotics | antiseptics | dead | live | red | toxins | white |

During vaccination, _____ or weakened pathogens are injected into the body.

This causes _____ blood cells to make _____ .

Later, when _____ pathogens enter the body, they are destroyed quickly. [4]

b) Vaccinations can provide immunity.

Give **one** other way by which a person can become immune to a pathogen. [1]

2 In March 2009, a nine-year-old girl was found to be infected with a new strain of the H1N1 swine flu virus. Over the next year many more people were found to have the swine flu virus.

The graph in **Figure 1** shows the number of reported cases of swine flu in the first 10 days of May 2010.

a) How many cases of swine flu had been reported by 5th May? [1]

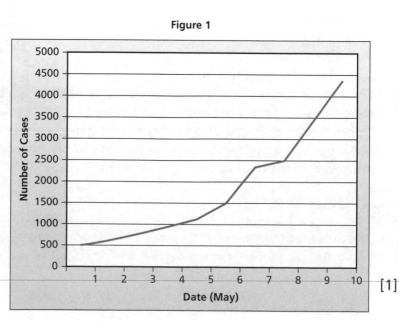

Figure 1

b) Which period showed the largest increase in the number of reported cases? [1]

c) Suggest why the spread of disease was so rapid. [2]

d) Why is it difficult to kill viruses inside the body? [2]

Total Marks .... / 11

Treating Diseases

1 Scientists often need to test bacteria for sensitivity to different antibiotics in order to decide the best antibiotic for treatment.
This is the method used:

1. Heat nutrient agar to 121°C.

2. Pour into a Petri dish and leave to cool.

3. Spread the bacteria onto nutrient agar in a Petri dish.

4. Place small filter-paper discs containing different antibiotics onto the agar.

5. Incubate the inoculated Petri dishes for 16 hours and then examine them.

a) What is nutrient agar? [2]

b) What is the main reason for heating the agar to 121°C? [1]
   Tick **one** box.

   So it pours easily ☐

   To dissolve the nutrients ☐

   To kill any microorganisms ☐

c) Why are the inoculated Petri dishes incubated before measurements are taken? [1]

Figure 1

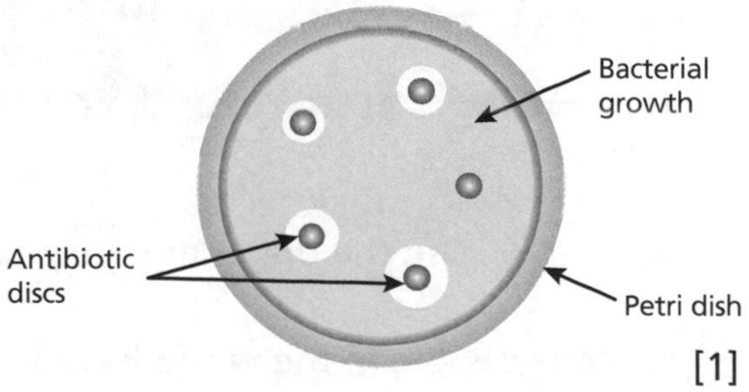

d) **Figure 1** shows the Petri dish after incubation.
   Erythromycin was found to be the most effective antibiotic.
   Penicillin had no effect at all on the bacteria.

Figure 1

   i) Label the antibiotic disc containing erythromycin with an **E**. [1]

   ii) Label the antibiotic disc containing penicillin with a **P**. [1]

2 HT **Figure 2** shows the stages of making a monoclonal antibody.

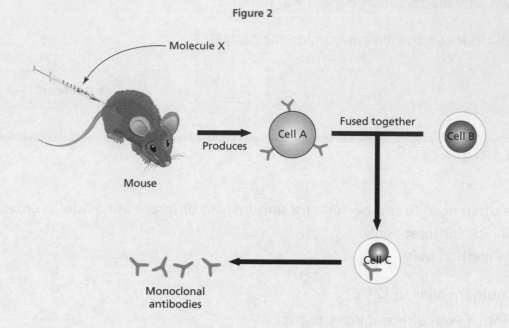

Figure 2

a) Identify the different types of cells labelled **A**, **B** and **C** in the diagram. [3]

b) What type of molecule is **X**? [1]
Tick **one** box.

Antibiotic ☐   Antigen ☐

Antibody ☐   Antiseptic ☐

c) Monoclonal antibodies can be used to treat some types of cancer.
This process is shown in **Figure 3**.

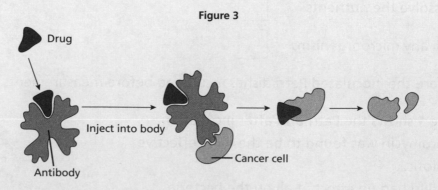

Figure 3

Use the diagram to explain how this process works. [4]

3 Many people in the world have a disease called arthritis.
Their joints become very painful.

Here is an article about a new arthritis drug.

To Use or Not to Use?

Arthritis is a very painful condition.

The problem with many drugs used to treat arthritis is that they have side effects.

One new drug was recently developed and tested on animals like mice and rats with no side effects. However, after a long-term study on human patients, side effects were noticed.

In this study, the drug was compared with a placebo. After 18 months of taking the drug, the risk of a patient having a heart attack was 15 out of 1000 compared with 7.5 out of 1000 for patients taking the placebo.

A decision has to be made about whether to use the drug even though it increases the risk of heart disease.

a) It is important that all drugs are tested on humans, and not just animals, before they are widely used.

Why is this? [1]

b) The long-term study used a placebo.

What is a placebo and why is it used? [3]

c) Some new drugs for arthritis are still allowed to be used even though they carry a slight risk.

Why do you think that this is? [2]

Total Marks _____ / 20

Plant Disease

1 Vinay is growing tomato plants in his garden.
He notices that the leaves of the plant have patches that are much lighter in colour.
He thinks that his plants may be infected by tobacco mosaic virus (TMV).

a) The presence of very light green patches on the leaves may mean that the plants will not grow as well.

Explain why this is. [3]

b) Vinay's gardening book says that the green patches might be caused by a lack of magnesium ions in the soil.

Why could this have an effect on the leaves? [1]

c) HT What could Vinay do to find out for certain if his plants had tobacco mosaic virus? [1]

Total Marks _____ / 5

Review Questions

Photosynthesis

1 Rebecca saw some pondweed producing bubbles in a tank.

Her teacher tells her that the bubbles contain oxygen.

Oxygen is made by the plant during photosynthesis.

a) Write down the word equation for photosynthesis. [2]

b) Rebecca investigates how quickly the pondweed produces oxygen.

Figure 1 shows the apparatus that she uses.

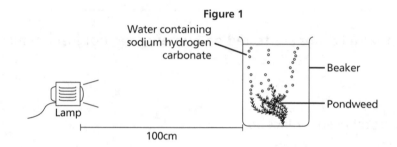

Figure 1

Rebecca adds different masses of sodium hydrogen carbonate to the water.

This gives the pondweed different concentrations of carbon dioxide to use for photosynthesis

She counts the number of bubbles given off each minute.

Table 1 shows the results.

Table 1

Mass of Sodium Hydrogen Carbonate Added in Grams	0.00	0.10	0.30	0.50	0.65	0.80
Number of Bubbles Given Off in 1 Minute	5	20	36	45	47	47

i) Describe the pattern shown by Rebecca's results. [2]

ii) Write down **two** ways in which Rebecca made sure that her results were valid. [2]

iii) The pondweed produces five bubbles of oxygen per minute without any sodium hydrogen carbonate being added to the beaker.

Why is the pondweed still able to produce some oxygen with no extra carbon dioxide? [1]

Total Marks _____ / 7

Respiration and Exercise

1 An athlete starts to exercise.

a) **i)** Aerobic respiration is taking place in her muscle cells.

Complete the balanced **word** equation for aerobic respiration.

glucose + _____ ⟶ _____ + water [2]

ii) The athlete's breathing rate increases as she continues to exercise.

Explain why. [3]

b) Towards the end of the exercise session, anaerobic respiration is taking place in the athlete's muscle cells.

Write the **word** equation for this type of anaerobic respiration. [1]

c) When the athlete has finished the exercise session, her breathing rate stays high for some time.

Explain why her breathing rate stays high after exercise. [3]

2 **Figure 1** shows one method of making wine.

a) Name the gas that bubbles up through the mixture. [1]

b) **i)** What type of respiration occurs during winemaking? [1]

ii) What substance gradually builds up in the wine as a result of this kind of respiration? [1]

c) Explain why sugar is used during winemaking. [2]

Figure 1

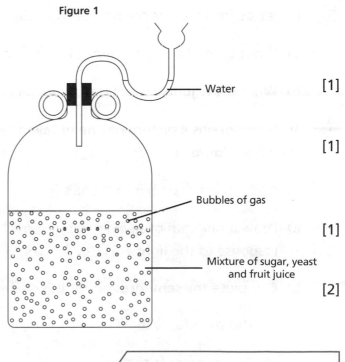

Water

Bubbles of gas

Mixture of sugar, yeast and fruit juice

Total Marks _____ / 14

Review Questions

Homeostasis and Body Temperature

1 **a)** Where are the receptors located that provide information about blood temperature? [1]
Tick **one** box.

Brain ☐

Lungs ☐

Kidneys ☐

Skin ☐

b) Circle the correct option to complete the sentence. [1]

The normal human body temperature is **20°C / 37°C / 75°C / 100°C**.

Total Marks .... / 2

The Nervous System and the Eye

1 Reflex actions are designed to prevent the body from being harmed.

a) What type of neurone carries a signal to the spinal cord in a reflex action? [1]

b) What is the junction between two neurones called? [1]

Figure 1

2 Matthew draws a picture of one of Jayna's eyes. His drawing is shown in **Figure 1**.

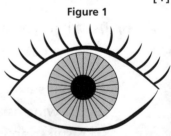

He then shines a torch into Jayna's eye.

a) Draw a diagram of Jayna's eye to show how it has changed in response to the light. [1]

b) Complete the sentences about this change.

i) The stimulus is the .... [1]

ii) The receptor is the .... [1]

iii) The response is carried out by the .... [1]

3 Figure 2 shows a section through the human eye.

Figure 2

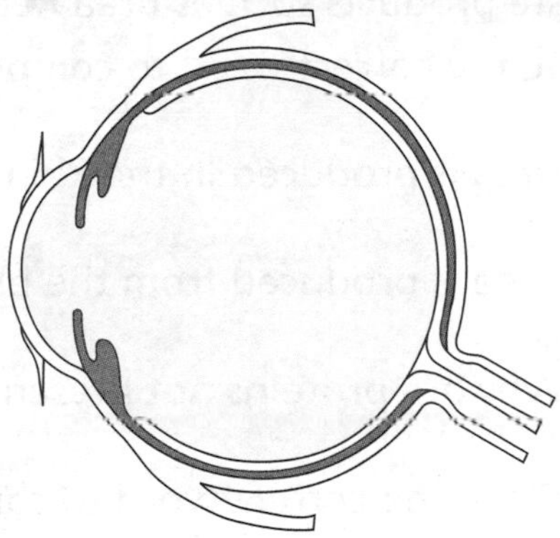

a) Add labels to the diagram to show the position of:

- The optic nerve. [1]

- The cornea. [1]

- The retina. [1]

b) Draw the lens and suspensory ligaments on the diagram. [2]

c) Describe the changes in the lens and suspensory ligaments when a person looks at a close object after looking at a distant object. [2]

d) Some people can only see close objects clearly and distant objects are out of focus.

Why is this and how can laser surgery be used to correct it? [2]

Total Marks _____ / 15

Hormones and Homeostasis

1 a) What does insulin cause glucose to be converted into? [1]

b) In which organ is this product mainly stored? [1]

c) **HT** What hormone does the body produce that converts this product back into glucose? [1]

d) **HT** Where is this second hormone produced? [1]

e) Which statement describes one of the causes of Type 1 diabetes?
Tick **one** box.

The pancreas does not produce insulin. ☐

The liver does not respond to insulin. ☐

The kidneys do not remove glucose from the blood. ☐

The liver does not produce insulin. ☐ [1]

2 Waste products such as urea need to be removed from the body.
Circle the correct word to complete each sentence.

    a) Urea is produced in the **kidneys / liver / lungs**.   [1]

    b) Urea is produced from the breakdown of **glucose / fats / amino acids**.   [1]

3   **a)** Why are proteins **not** present in the urine of a healthy person?   [1]

    b) Circle the correct word to complete the sentence.

        After filtration, all the glucose is **reabsorbed / respired / released**.   [1]

4 A scientist carries out an experiment to see how much sweat and urine a person produces at various temperatures.

    Table 1 shows the results for sweat production.

Table 1

Air Temperature (°C)	Sweat Produced (cm³ per hour)
0	2
5	4
10	8
15	16
20	40
25	64
30	110
40	200

The scientist produces a graph to show the amount of urine produced at each temperature.

Figure 1 shows this graph.

a) On the scientist's graph:

- Plot the data from **Table 1** that shows sweat production at each temperature.

- Finish the graph by drawing the best curve through the points.

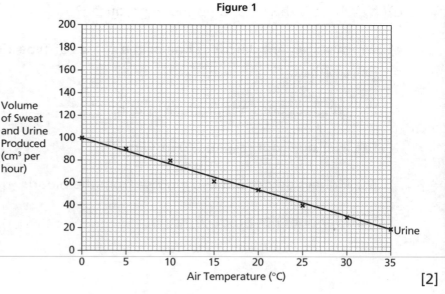

Figure 1

    [2]

b) Compare the changes in the volume of sweat and urine produced as the temperature increases. [3]

c) HT Explain why the volumes of sweat and urine change with temperature. [4]

Total Marks .... / 18

Hormones and Reproduction

1 The female menstrual cycle is controlled by hormones, which cause eggs to be released and bring about changes to the uterus lining.

a) Which hormone is secreted from the pituitary gland and causes eggs to mature in the ovaries? [1]

b) Which hormone is released from the ovaries and causes the uterus lining to thicken? [1]

c) Explain how hormones given to women can:

i) HT Increase fertility? [2]

ii) Reduce fertility? [2]

Total Marks .... / 6

Plant Hormones

1 Which of the following statements about plants are true?
Tick all **true** statements.

A Plant shoots grow towards light. ☐

B Plant shoots grow towards moisture. ☐

C Roots contain special cells called auxin cells. ☐

D Roots grow in the direction of gravity. ☐

E Plant auxins can be used as weedkillers. ☐

F Plant auxins can be used as medicine. ☐

G Gravitropism is a response by the plant to the force of gravity. ☐ [4]

Total Marks .... / 4

Practice Questions

Sexual and Asexual Reproduction

1 **a)** Use the words in the box to label the cell in **Figure 1**.

chromosomes	nucleus	cytoplasm	cell membrane

Figure 1

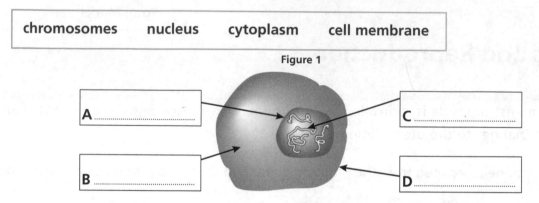

A ....

B ....

C ....

D ....

[4]

b) Choose the correct word to complete each sentence.

Sexual reproduction is the **division / separation / fusion** of the male and female gametes.

The resulting offspring will contain **DNA / cells / enzymes** from both parents.

This gives rise to **fertilisation / differentiation / variation**. [3]

2 **Figure 2** shows the sequence of events used to clone sheep by embryo transplantation.

Figure 2

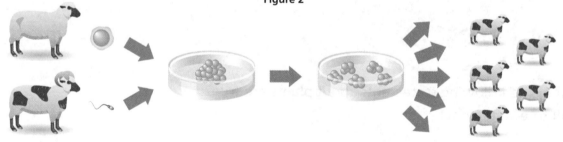

a) Use words from the box to complete the sentences.

gametes	old	implanted	shocked	directed	characteristics
sexual	wombs	asexual	sterile	stomachs	specialised

A male and female sheep with the desired .... are mated using IVF.

This is an example of .... reproduction.

Before the cells become ...., the embryo is split into several clumps.

These are then .... into the .... of surrogate sheep. All the resulting offspring are identical to each other. [5]

b) Why would farmers want to use embryo transplants rather than more natural methods of reproduction? [2]

c) Explain why the offspring from the above process will be identical to each other but not identical to the original male and female sheep. [4]

d) Scientists now have the technology to clone human embryos.

Give **one** medical reason why cloning human embryos might be permitted. [1]

Total Marks .... / 19

DNA and Protein Synthesis

1 The genetic material in the nucleus of a cell is made from DNA.

a) In which structures is the DNA contained? [1]

b) In terms of DNA, describe what a gene is and what it does. [3]

c) What is the name given to the shape of a DNA molecule? [1]

d) HT In terms of DNA, what is a mutation? [1]

Total Marks .... / 6

Patterns of Inheritance

1 a) Circle the correct pairs of human sex chromosomes.

XY and YY   XX and XY   XX and YY   XF and XM

[1]

b) In the space to the right, draw a genetic cross diagram to show how these chromosomes are involved in sex inheritance. [2]

2 Draw **one** line from each definition to the correct genetic term.

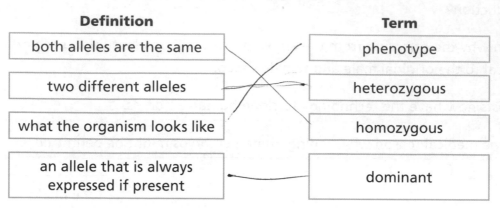

Definition	Term
both alleles are the same	phenotype
two different alleles	heterozygous
what the organism looks like	homozygous
an allele that is always expressed if present	dominant

[3]

3 Fruit flies are often used in genetic crosses.

There are two types of wings on fruit flies: short or normal.

Answer the following questions.

Use the letter **N** to represent normal wings and **n** to represent short wings.

a) What is the phenotype for a fly that has the homozygous dominant genotype? [1]

b) A heterozygous male fly mates with a homozygous recessive female.

Complete the diagram in **Figure 2** for this genetic cross.

c) What will be the ratio of normal wings to short wings in the offspring?

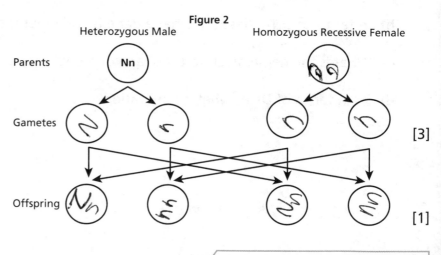

Figure 2

[3]

[1]

Total Marks _____ / 11

Variation and Evolution

1 It is thought that many years ago members of the giraffe family had short necks.

Giraffes now have longer necks, which allow them to reach food higher in the trees.

a) Use Charles Darwin's theory of natural selection to explain how modern giraffes with longer necks may have evolved.

[3]

b) Lamarck was another scientist with ideas about evolution.
He had a theory that some giraffes grew longer necks in order to reach the leaves high on the trees.
These giraffes were then more successful and were able to breed and pass their long necks onto their offspring.

Explain why Lamarck's theory is **not** correct. [2]

c) Darwin also suggested that humans and apes evolved from a common ancestor.

Give **two** reasons why Darwin's theories were not accepted by some people. [2]

Total Marks _____ / 7

Manipulating Genes

1 HT **Figure 1** shows the **first** stage in the process of insulin production using genetic engineering.

Figure 1

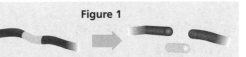

a) What do scientists use to 'cut out' the insulin gene from the chromosome? [1]

b) The 'cut' gene is then inserted into a bacterium.

Why are bacteria good host cells for the 'cut' insulin gene? [2]

Total Marks _____ / 3

Classification

1 **a)** Describe **one** way in which fossils are formed. [1]

b) Explain why fossils can be quite hard to find. [1]

c) Many fossils are of animals that are extinct.

Give **three** factors that could contribute to the extinction of a species. [3]

d) Give an example of a **species** that is now extinct. [1]

Total Marks _____ / 6

Practice Questions

Ecosystems

1 A number of animals live in the Sahara desert.

a) Suggest **two** major problems that animals living in the desert have to deal with. [2]

b) The cactus is a plant that is adapted to survive in desert environments.

Suggest how the following adaptations help the cactus to survive:

i) The cactus has a thick stem. [1]

ii) The cactus has spines instead of leaves. [1]

2 Gurjot and Ben want to investigate the effect of animals grazing on the numbers and types of plants in an area.
They get permission from the owner of a field grazed by horses and use a square metre quadrat.

a) Gurjot suggests they place five quadrats in the field as shown in **Figure 1**.

Ben says that this method might not give a very accurate estimate of the plants growing in the field.

Is Ben correct?
You must give a reason for your answer. [3]

Figure 1

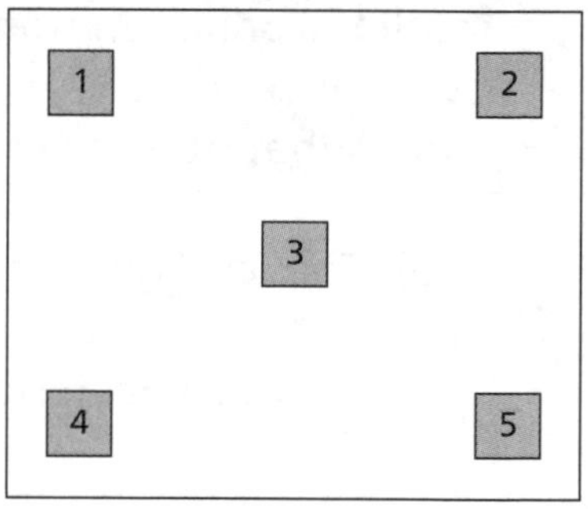

b) Gurjot and Ben count the numbers of dandelion, dock and thistle plants inside each quadrat.
They calculate the mean number of each type of plant per square metre.
They repeat the experiment in another field grazed by sheep and a third field that has no animals grazing.

The results are shown in **Figure 2**.

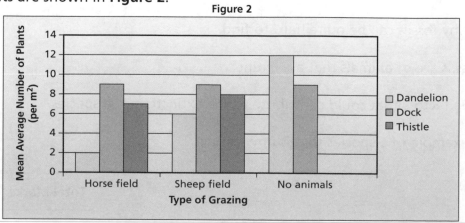

Figure 2

The number of thistles in the five quadrats in the field with no animals is:
5, 7, 9, 4 and 10.

Use this information to plot the mean result on the chart in **Figure 2**. [3]

c) The number of docks in the five quadrats in the field with no animals was 12, 6, 9, 7 and 11. What was the median number of docks in this field? [1]

d) Which field has the least number of dandelions? [1]

e) If the sheep field has an area of 1000 square metres, calculate how many dandelion plants there are in the whole field. You must show your working. [2]

f) What conclusions about the effect of animals grazing on plants could you draw from these results? [4]

g) Suggest **one** way in which Ben and Gurjot could improve the reliability of their results. [1]

h) Name **three** things that the dandelions, docks and thistles will compete with each other for. [3]

Total Marks _____ / 22

Cycles and Feeding Relationships

1 In an area of marshland there are numerous plants that are eaten by insects.
The insects are eaten by frogs.
Herons eat the frogs.

Figure 1

a) Write the names of the organisms in the correct place on **Figure 1** to complete the pyramid of biomass. [3]

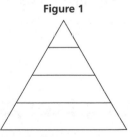

b) What is the source of energy for all of the organisms in the marshland? [1]

c) Why is the total biomass at the top of the pyramid much less than the total biomass at the base of the pyramid? [1]

d) A farmer sprays a nearby field with pesticide.
Some of the spray falls on the marshland and kills the insects.

How will this affect the number of frogs? Give a reason for your answer. [2]

2 **Figure 2** shows some parts of the carbon cycle.

Figure 2

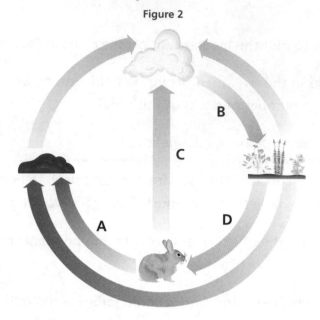

a) Choose the letter from **Figure 2** that corresponds with each process.

   i) Egestion   [1]

   ii) Feeding   [1]

   iii) Photosynthesis   [1]

   iv) Respiration   [1]

b) Name another process, not shown on the diagram, which releases carbon dioxide
into the air.   [1]

> Total Marks _____ / 12

Disrupting Ecosystems

1 Circle the correct options to complete the sentences.

When deforestation occurs in **tropical / arctic / desert** regions, it has a devastating impact
on the environment.

The loss of **trees / animals / insects** means that less photosynthesis takes place, so less
oxygen / nitrogen / carbon dioxide is removed from the atmosphere.

It also leads to a reduction in **variation / biodiversity / mutation**, because some species may become **devolved / damaged / extinct** as **habitats / land / farms** are destroyed. [6]

Feeding the World

1 **Figure 1** shows how the energy supplied to a cow is transferred.

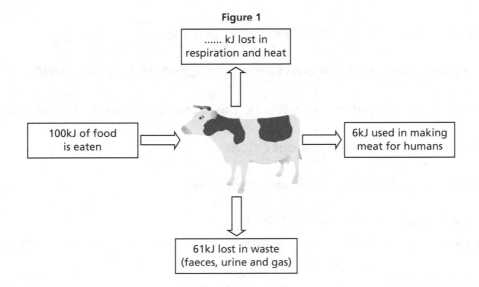

Figure 1

a) Calculate how much energy is transferred from the cow in respiration and heat.
   You must show your working. [2]

b) Calculate the total amount of energy transferred from the cow to the environment.
   You must show your working. [2]

c) How can the efficiency of meat production in cows be improved?
   Tick **two** boxes [2]

Limit the movement of the animals. ☐

Heat the animals' surroundings. ☐

Allow plenty of space for the animals to move. ☐

Keep the animals' surroundings cool. ☐

Review Questions

Sexual and Asexual Reproduction

1 Spider plants reproduce asexually by producing structures called stolons.

This is shown in **Figure 1**.

Circle the correct word to complete each sentence.

Figure 1

Spider plant stolons

Stolon – a rooting side branch   New individual established   Now independent

a) Asexual reproduction needs **one / two** spider plant(s). [1]

b) Asexual reproduction does not involve the production of **gametes / DNA**. [1]

c) The genes in the new spider plants will be **the same as / different to** the parent genes. [1]

d) The new spider plant is called a **shoot / clone**. [1]

2 Mitosis is the division of body cells to make new cells.

a) When is mitosis **not** used for cell division?
Tick **one** box.

Asexual reproduction ☐

Gamete production ☐

Repair ☐

Growth ☐ [1]

b) Complete the sentences about mitosis.

A copy of each _____ is made before a cell divides.

The new cell has the same _____ information as the _____ cell.

Meiosis takes place in the testes, and produces sperm containing 23 _____. [4]

c) What type of cell is produced in meiosis? [1]

3 The statements below describe the different stages in adult cell cloning.
They are in the wrong order.

a) Number the statements **1** to **6** to show the correct order.
Two have been done for you.

The nucleus is removed from an egg cell and discarded.

The embryo is implanted in the womb of another cow.

The nucleus of the body cell is placed into the empty egg cell.

A body cell is taken from a prize bull.

The cell is zapped with electricity and starts to divide to form an embryo. **5**

An unfertilised egg cell is taken from a cow. **2** [3]

b) At stage 5, how are the cells prompted to divide? [1]

Total Marks _____ / 14

DNA and Protein Synthesis

1 HT Explain how genes control protein synthesis. [4]

2 a) Define the term 'genome'. [1]

b) Suggest **two** important uses of this data about the human genome. [2]

Total Marks _____ / 7

Patterns of Inheritance

1 History tells how King Henry VIII was so desperate to have a male heir that he divorced or disposed of all his wives who were unable to produce a son.

a) What sex chromosomes are found in eggs? [1]

b) What sex chromosomes are found in sperm? [1]

c) Use your answers to part a) and b) to explain why it was unfair of Henry to blame his wives for the lack of a son. [2]

2 Rita likes to grow plants.

One particular plant she grows can either have red flowers or white flowers.

She decides to cross a red flowered plant with a white flowered plant.

a) Complete the table below to show this genetic cross.

Use **R** to represent the dominant allele and **r** to represent the recessive allele. [4]

		White Flower (rr)	
		Genotype of Ovum (r)	Genotype of Ovum (r)
Red Flower (Rr)	Genotype of Pollen (R)		
	Genotype of Pollen (r)		

b) Rita grows 24 plants using the seeds from this cross.

Predict the number of red flowered and white flowered plants that are produced. [2]

c) Why is it unlikely that the actual numbers of each type of plant will exactly match this prediction? [2]

Total Marks _____ / 12

Variation and Evolution

1 Variation can be due to inherited factors, environmental factors or a combination of both.

Cathy and Drew are sister and brother.
Table 1 shows how they are different.

Table 1

Cathy	Drew	Inherited (I), Environmental (E) or Both (B)
tongue roller	non roller	
not colour blind	colour blind	
1.5m tall	1.6m tall	
speaks French	does not speak French	

Complete **Table 1** to show if each difference is caused by inherited factors (**I**), environmental factors (**E**) or a combination of both (**B**). [4]

Total Marks _____ / 4

Manipulating Genes

1 Farmers have been using selective breeding for thousands of years to produce crops and animals with desirable characteristics.

a) Suggest **two** characteristics that might be desirable in food crops. [2]

b) Suggest two characteristics that might be desirable in dairy cows. [2]

c) Describe the main stages in the process of selective breeding. [4]

d) Developments in biotechnology mean that genes can now be inserted into crop plants to give them desirable characteristics.

What is the term used to describe crops that have had their genes altered in this way? [1]

Total Marks _____ / 9

Classification

1 Scientists frequently study the distribution of the snail, *Cepaea nemoralis*. The snail has a shell that can be brown or yellow.

a) i) What genus does the snail belong to? [1]

ii) How could scientists prove that the different coloured snails were all the same species? [2]

b) Scientists believe that shell colour affects the body temperature of the snails. Snails with dark shells warm up faster than those with light shells. In cold areas, this would be advantageous to the dark-coloured snail.

The average annual temperature in Scotland is 2°C lower than in England.

Use the theory of natural selection to explain why populations of the snail in Scotland contain a higher percentage of dark-coloured snails? [3]

c) Explain how two new species of snails could be formed from two different populations of *Cepaea*. [4]

Total Marks _____ / 10

Review Questions

Ecosystems

1 **Figure 1** shows some adaptations of the polar bear.

Figure 1

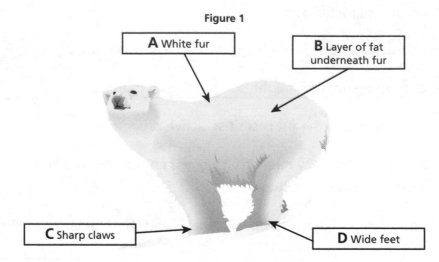

A White fur

B Layer of fat underneath fur

C Sharp claws

D Wide feet

Match the adaptations to their functions by writing the letter **A**, **B**, **C** or **D** alongside each function.

You can use each letter more than once or not at all.

a) For catching seals. [1]

b) For warmth. [1]

c) To help grip the ice. [1]

d) For camouflage. [1]

2 A class of students was asked to estimate the number of daisies on the school football pitch. The pitch is 60m by 90m.

They decided to use quadrats that were 1m².

a) Which is the best way of using quadrats in this investigation?
Tick **one** box.

Place all the quadrats where there are lots of plants. ☐

Place all the quadrats randomly in the field. ☐

Place all the quadrats where daisies do not grow. ☐ [1]

b) Each student collected data by using 10 quadrats.

Table 1 shows the results of one student, Shaun.

Table 1

Quadrat	1	2	3	4	5	6	7	8	9	10
Number of Daisies	5	2	1	0	4	5	2	0	6	3

Calculate the mean number of daisies per quadrat counted by Shaun.
You must show your working. [2]

c) Another student, Mandeep, calculated a mean of 2.3 daisies per quadrat from her results.

i) Use Mandeep's results to estimate the total number of daisies in the whole pitch.
You must show your working. [2]

ii) The centre circle has a diameter of 10 metres.

How many daisies are likely to be in the centre circle? (Use π = 3.14) [3]

Total Marks _____ / 12

Cycles and Feeding Relationships

1 Underline the correct words to complete the sentences about the carbon cycle.

Plants and **animals / algae** remove carbon dioxide from the air.

When plants die, they are broken down by **consumers / decomposers / producers**.

Bacteria and fungi are examples of **consumers / decomposers / producers**. [3]

Total Marks _____ / 3

Disrupting Ecosystems

1 **a)** Complete the sentences.

Some gases in the atmosphere prevent _____ from escaping into space.

This is called the _____ effect.

Two gases that contribute to this are _____ and _____. [4]

b) Which of the following are possible negative effects of global warming?
Tick **two** boxes.

Climate change ☐   A rise in sea level ☐

Erosion of buildings ☐   Increase in available land ☐

Deforestation ☐   [2]

2 Some students investigate the effect of different concentrations of herbicide on weeds.
They divide a plot of land into four equal sections.
Table 1 shows the concentration of herbicide used in each section of the field.

Table 1

Section A	Section B	Section C	Section D
Sprayed with 1% concentration of herbicide	Sprayed with 2% concentration of herbicide	Sprayed with 4% concentration of herbicide	Sprayed with 8% concentration of herbicide

After a week, the students count the number of healthy weeds and dead weeds on each section.
The pie charts in **Figure 1** show their results.

Figure 1

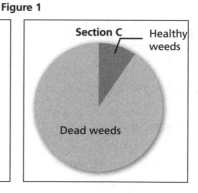
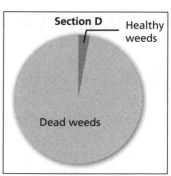

a) Approximately what percentage of weeds is killed by the herbicide of 1% concentration?   [1]

b) What concentration of herbicide would you recommend to kill weeds?
Explain your answer.   [3]

c) Suggest **two** factors that the students need to control in their investigation.   [2]

d) Give a reason why weedkillers should not be used:

   i) To kill weeds growing at the edge of a pond.   [1]

   ii) To kill weeds in a hedgerow in the countryside.   [1]

Total Marks _____ / 14

Feeding the World

1 **Figure 1** shows energy transfer in photosynthesis.

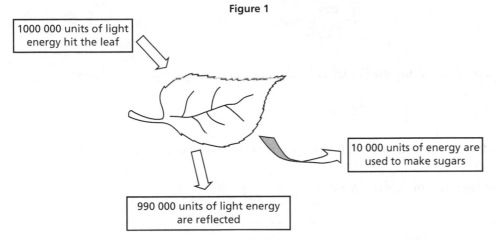

Figure 1

1000 000 units of light energy hit the leaf

990 000 units of light energy are reflected

10 000 units of energy are used to make sugars

a) What percentage of the light energy hitting the leaf is trapped by photosynthesis? [1]

b) A squirrel eats the leaf.
   Figure 2 shows the energy intake and output of the squirrel.

Figure 2

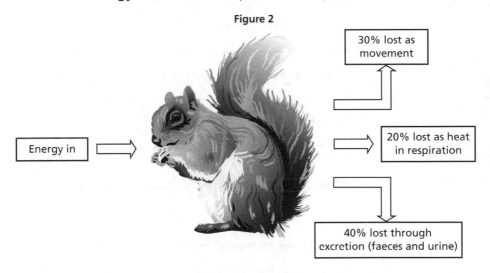

30% lost as movement

Energy in

20% lost as heat in respiration

40% lost through excretion (faeces and urine)

What percentage of the squirrel's energy intake is used for growth? [2]

c) Use your answers from parts **a)** and **b)** to calculate what percentage of light energy hitting the leaf is used by the squirrel for growth.
   You must show your working. [2]

d) Suggest why it is more energy efficient to have a vegetarian diet. [3]

Total Marks _____ / 8

Mixed Exam-Style Questions

1 **Figure 1** shows a cheek cell.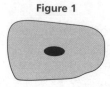

Figure 1

a) Put a ring around the most likely width of this cell.

   0.002mm   0.02 mm   2mm   20mm   [1]

b) Write down **three** structures that are found in plant cells but **not** in animal cells.   [3]

c) Use words from the box to complete **Table 1**.

chloroplast   cell membrane   nucleus   vacuole

   Table 1

Sub-Cellular Structure	Function
	stores cell sap
	allows gases and water to enter and leave the cell
	controls what the cell does

   [3]

2 **Figure 2** shows the villi that line the small intestines in a healthy person.

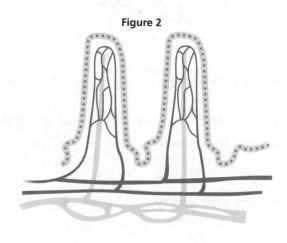

Figure 2

a) Describe **two** features of the villi shown in **Figure 2** that help the small intestine to function efficiently. [2]

b) The villi that line the small intestine are also covered in microvilli.
Each cell that has microvilli is packed with lots of mitochondria.

   i) Explain the advantage of microvilli in the absorption of digested food molecules. [2]

   ii) Explain why it is necessary for these cells to be packed with lots of mitochondria. [2]

c) The villi of a person with coeliac disease are damaged and have a much smaller surface area compared with the villi of a healthy person.

What effect will this damage have on the function of the small intestine? [1]

3 A sports scientist investigated the amount of lactic acid in the leg muscle of a runner.

The runner ran for 30 minutes and then rested.

Table 2 shows the results.

Table 2

Time After Start of Exercise (Minutes)	0	10	20	30	40	50	60	70	80	90
Lactic Acid (Arbitrary Units)	0	1	6	13	8	6	4	3	1	0

a) What was the percentage increase in lactic acid between 20 and 30 minutes? [2]

b) How long after the exercise finished did it take for the lactic acid to be completely removed from the muscle? [1]

c) HT Explain what happens to the lactic acid build-up after exercise. [2]

d) The lactic acid of a second athlete is investigated in the same way.

Why is it important to keep variables the same in the investigation? [1]

e) It takes 20 minutes for the lactic acid levels in the second athlete's muscle to return normal.

Which athlete is the fittest? [1]

4   This article about squirrels was published in a natural history magazine.

Red Versus Grey?

There are two main species of squirrel living in Britain: the native red squirrel (*Sciurus vulgaris*) and the grey squirrel (*Sciurus carolinensis*), which was introduced from America.

The number of reds has declined in Britain over the last sixty years and they are now rare.

There have been a number of studies to try to find out why the number of reds has declined.

The main reason is probably due to competition. The reds are lighter animals and so spend more time up in the trees. They prefer coniferous woodland, feeding on the seeds from pine cones, high up in the trees. The larger greys spend more time on the ground and are better adapted to eating acorns and other seeds that are found in the more common, deciduous woodlands.

The reds lack the enzymes to digest acorns. Therefore, they are soon out-competed in deciduous woodlands.

a)   What information in the article shows that the red and grey squirrels are closely related?   [1]

b)   The reason for the decline of the red squirrels is thought to be competition.

     What are the squirrels competing for?   [1]

c)   Why are the two types of squirrel **not** competing for mates?   [1]

d)   Some conservationists want to put up boxes that will give out poisoned bait to any squirrels that are heavier than a certain mass.

     Suggest **one** advantage and **one** disadvantage of this conservation method.   [2]

e)   Conservationists hope that a new strain of the red squirrel will evolve, which can digest acorns.

     Explain how natural selection could bring this about.   [4]

5 HT A plant is receiving plenty of light but the rate of photosynthesis stops increasing.

What other **environmental** factors might be responsible?
Tick **one** box.

Concentration of carbon dioxide or the amount of oxygen ☐

Concentration of carbon dioxide or the temperature ☐

Concentration of chlorophyll or the temperature ☐

Concentration of glucose or the amount of oxygen ☐ [1]

6 A student germinated a bean seed.
In **Figure 3**, **A** shows the bean after seven days.
The student then turned the bean onto its side, as shown by **B**.
C shows the bean a week later.

Figure 3

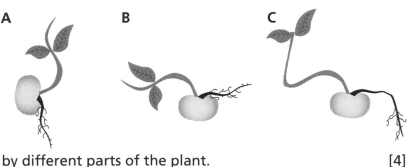

Describe the type of responses shown by different parts of the plant. [4]

7 **Figure 4** shows the inheritance of cystic fibrosis in a family.

Figure 4

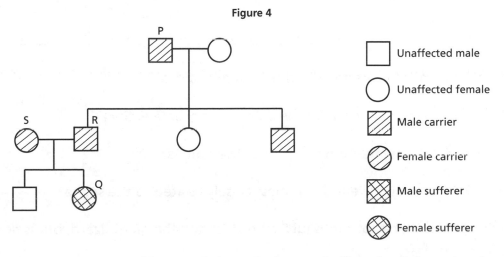

Unaffected male

Unaffected female

Male carrier

Female carrier

Male sufferer

Female sufferer

a) Using **C** for the non-cystic fibrosis allele and **c** for cystic fibrosis allele, write down the genotype of:

i) Person **P**. [1]

ii) Person **Q**. [1]

b) What is the phenotype of ~~per~~son **Q**? [1]

c) Draw a genetic diagram and work out the percentage probability of a carrier and a sufferer having a baby that has cystic fibrosis. [4]

8 **Figure 5** shows an evolutionary tree for some present-day vertebrates.

Figure 5

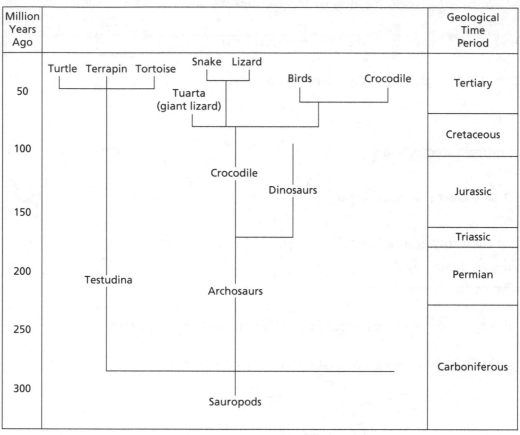

a) How many millions of years ago did the *Testudina* appear? [1]

b) In what geological time period did the dinosaurs become extinct? [1]

c) How do scientists know that dinosaurs once lived on Earth? [1]

d) What group of animals alive today is most closely related to the snake? [1]

e) Which ancestor is shared by dinosaurs, crocodiles and the giant lizard, but is not an ancestor of tortoises? [1]

9   The MMR vaccination is usually given to children when they are 13 months old.

It protects them from three diseases: measles, mumps and rubella.

a) Explain how a vaccination like MMR can protect a person from getting a disease.   [4]

b) MMR contains three different vaccinations in one injection.
About 10 days after the injection, the child might develop a fever and a measles-like rash.
After three weeks, the child might get a very mild form of mumps.
After six weeks, a rash of small spots like rubella may develop.
However, most children suffer no symptoms after vaccinations.

   i)   People may feel ill after having a vaccination.

   Explain why this is.   [2]

   ii)   Some parents do not want their children to have the MMR vaccine.
   They would like their children to visit the doctor three different times for three
   separate vaccinations.

   Suggest why they might feel this way.   [2]

   iii) Suggest why the government might want children to have one vaccination rather
   than three separate injections.   [1]

c) Doctors are worried that the number of children being vaccinated might fall.

Explain why it is important that the percentage of children vaccinated stays above
a certain level.   [2]

10   **Figure 6** shows a red-eyed tree frog.
These frogs are bright green with red eyes, blue stripes and
orange feet.
They live in trees in the rainforest and feed on small insects.

Figure 6

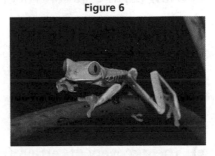

Explain how each of the following adaptations helps the frog
to survive:

a) It has sticky pads on its 'fingers' and toes.   [1]

b) When sleeping, it hides its bright colours by closing its eyes and tucking its feet beneath its body.   [1]

c) It has a long, sticky tongue.   [1]

Mixed Exam-Style Questions

11 There are a number of theories to explain the origin of all the living organisms that exist today.

The information below describes three of these theories.

Charles Darwin's Theory

Darwin's theory states that all organisms are slightly different. The organisms that are best suited to their environment will pass on their characteristics and so the population will gradually change.

Lamarck's Theory

Lamarck's theory states that organisms are changed by their environment during their life. They then pass on their new, acquired characteristics and so the population will change.

New Earth Creation Theory

The creation theory states that the Earth, and life on it, was created by God, less than 10 000 years ago. Only very minor changes within various species have happened since creation.

a) Which one of these theories involves natural selection?
   Tick **one** box.

   Charles Darwin's Theory ☐

   Lamarck's Theory ☐

   New Earth Creation Theory ☐   [1]

b) HT Charles Darwin did not know why organisms show variation.
   Scientists now know that mutations play a role in variation.

   What is a mutation?   [1]

c) The discovery of certain fossils has made some people reject the New Earth Creation Theory.

   Explain why.   [2]

12 Duckweed is a small floating plant found in ponds.
It reproduces quite quickly to produce large populations.

Helen decides to investigate duckweed by growing some in a beaker of water.
She counts the number of duckweed plants at regular intervals.

Table 3 shows her results.

Table 3

Day	1	4	8	10	15	19	20	23	26	30
Number of Plants	1	2	3	4	15	28	28	28	29	29

a) Plot a graph of the results to show how the population changed over time.

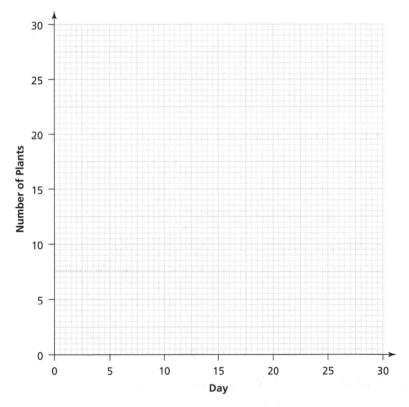

[3]

b) Describe what is happening to the population between day one and day ten. [1]

c) Suggest **two** factors that could have prevented the population from continuing
to increase. [2]

d) Explain how Helen could use this experiment to compare the amount of fertiliser
that has been washed into two different ponds. [3]

Mixed Exam-Style Questions

13 There are a number of ways in which we can help the environment.

a) Draw one line from each method of helping the environment to the corresponding result.

Method	Result
Increasing the amount of metal recycled	Less acid rain is produced
Reducing acid emissions	Fewer quarries are dug to provide raw materials
Using fewer pesticides and fertilisers	Fewer forests are cut down
Increasing the amount of paper recycled	Less pollution of rivers flowing through farmland

[3]

b) In the past, bags of compost from garden centres contained lots of peat.

Explain how this can damage the environment. [2]

c) Sustainable food production is another way of helping the environment.

Explain what is meant by 'sustainable food production'. [2]

14 a) How many chromosomes does a human body cell contain? [1]

b) How may chromosomes do human sex cells contain?
Tick **one** box.

Half the number of chromosomes in a normal body cell. ☐

The same number of chromosomes in a normal body cell. ☐

Half the number of chromosomes in a sperm cell. ☐

Twice the number of chromosomes in a normal body cell. ☐ [1]

c) Explain what determines the sex of an individual. [2]

d) The probability of having a baby boy is 50%.

Draw a genetic diagram to show why this is. [3]

15 **Figure 7** shows a special type of fermenter that is used to grow mycoprotein.

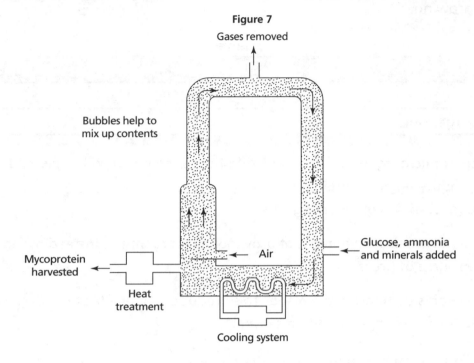

Figure 7

Gases removed

Bubbles help to mix up contents

Glucose, ammonia and minerals added

Air

Mycoprotein harvested

Heat treatment

Cooling system

a) What type of microorganism is grown in the fermenter in order to produce mycoprotein? Circle **one** answer.

   bacterium　　　　　　**fungus**　　　　　　protist　　　　　　virus　　　[1]

b) Air is bubbled into the fermenter.

   i) What do the microorganisms need the air for? [2]

   ii) The air helps to circulate the mixture.
   This is used instead of a metal stirrer, which would break up the long strands of mycoprotein that are made.

   How does using this mixing process help to produce a good meat substitute? [1]

c) **Table 4** shows how long it takes for different organisms to double their protein content when growing.

Table 4

Organism	Number of Hours Taken to Double Protein Content
cattle	4000
microorganisms	8

i) If 2g of microorganisms were added to the fermenter, how long would it take to produce 64g of protein?
You must show your working. [2]

ii) Write down **two** advantages of growing microorganisms, instead of farming cattle, as a source of protein. [2]

16 The kidney works by filtering the **blood**. This produces kidney **filtrate**.
The filtrate is then changed and forms **urine**.

Table 5 shows the composition of these three fluids.

Table 5

Substance	Contents (g per 100cm³)		
	Blood Plasma	Kidney Filtrate	Urine
protein	6.80	0.00	0.00
glucose	0.10	0.10	0.00
urea	0.02	0.02	2.00

a) Which **two** substances are found in blood plasma but not in urine? [2]

b) How many times more concentrated is the urea in the urine compared with the urea in the blood plasma?
You must show your working. [2]

c) The filter units in the kidney only allow small molecules to pass through into the kidney filtrate.

i) Why is glucose found in the kidney filtrate but not proteins? [1]

ii) Explain why there is usually no glucose in the urine. [2]

17 Camels are adapted to living in hot, sandy deserts.

a) Explain why it is important for mammals to keep their body temperature fairly constant. [2]

b) Different sized mammals have to lose different amounts of water to keep their body temperature constant in the desert.

This is shown in **Figure 8**.

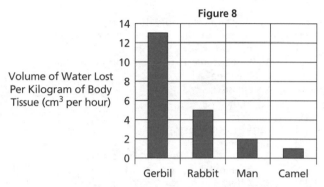

i) Describe the effect of increasing body size on the amount of water lost per kilogram of tissue. [1]

ii) A camel weighs 500 kg.

Use the bar chart to calculate the volume of water this camel would have to lose in one hour to keep its body temperature constant. [2]

iii) Humans must maintain a constant body temperature.
A camel's body temperature can rise by about 7°C without it coming to any harm.

Explain how this may be an advantage to the camel in the desert. [2]

> **Total Marks** .... / 110

Answers

Pages 6–7 Review Questions

1. Cells [1]
2. a) Cell membrane [1]; cytoplasm [1]
   b) **Any two of:** cell wall [1]; chloroplast [1]; vacuole [1]
3. a) cells [1]
   b) Tail [1]
   c) Testis [1]
   d) Fertilisation [1]
4. Carbon monoxide: binds with haemoglobin / less oxygen carried in blood [1]; Nicotine: addictive / raises blood pressure [1]; Tar: causes lung cancer / bronchitis [1]
5. Photosynthesis [1]
6. (Aerobic) respiration [1]
7. **Any two of:** supports the body [1]; acts as a framework that enables muscles to move the body [1]; protects the organs [1]; makes red blood cells [1]
8. a) i) C [1]
      ii) D [1]
      iii) A [1]
      iv) D [1]
   b) An organism that feeds on / in another living organism [1]
9. a) Bacteria / fungi / virus / protista [1]
   b) To prevent harmful microorganisms from entering the medic's body [1]
10. a) i) Oxygen [1]
       ii) Carbon dioxide [1]
    b) To allow gases to pass across [1]
    c) To increase the rate at which gases can leave or enter the blood [1]

Pages 8–25 Revise Questions

Page 9 Quick Test
1. Nucleus
2. On the ribosomes
3. Cell wall; chloroplasts; permanent vacuole
4. To support the cell
5. In loops / plasmids in the cytoplasm, floating free

Page 11 Quick Test
1. Liver cell, nucleus, bacterium, ribosome
2. 5μm
3. $\frac{50}{0.025}$ = 2000 times
4. **Any two of:** sterilise the agar first; use sterile loop; tape lid so that it does not come off

Page 13 Quick Test
1. 46 (23 pairs)
2. So that each new cell gets the full amount of chromosomes
3. A cell that is not specialised and can divide to form various types of cells
4. To produce new cells so the root and shoot can grow

Page 15 Quick Test
1. The perfume diffuses faster

2. Energy from respiration
3. Thin walls / rich blood supply

Page 17 Quick Test
1. To become specialised and more efficient
2. To provide energy for the sperm to swim
3. A tissue
4. To cover the body
5. An organ

Page 19 Quick Test
1. A protein
2. **Any two of:** temperature; pH; concentration of enzyme; concentration of substrate
3. In the stomach, pancreas and small intestine
4. Lipase
5. In the liver

Page 21 Quick Test
1. Platelets
2. No nucleus; biconcave disc; contain haemoglobin
3. Artery
4. Right atrium
5. Valves

Page 23 Quick Test
1. a) Communicable
   b) Non-communicable
   c) Non-communicable
2. The UV light can trigger cancer
3. Being overweight
4. Layers of fatty material build-up in the arteries, reducing the blood flow so that not enough oxygen and glucose (needed for respiration) can reach the heart muscle

Page 25 Quick Test
1. Palisade mesophyll
2. Water and minerals
3. **Any two of:** low light; high humidity; low temperature; still air
4. To stop the plant losing water when carbon dioxide is not needed

Pages 26–33 Practice Questions

Page 26 Cell Structure
1. 1 mark for each correct row [3]

Type of Cell	Nucleus	Cytoplasm	Cell Membrane	Cell Wall
Plant Cell	✓	✓	✓	✓
Bacterial Cell	✗	✓	✓	✓
Animal Cell	✓	✓	✓	✗

Page 26 Investigating Cells
1. a) $\frac{50}{0.1}$ [1]; = 500 times [1]
   b) i) E [1]
      ii) A [1]
      iii) C [1]
      iv) B [1]
      v) A [1]
      vi) D [1]
      vii) C [1]
      viii) B [1]
2. a) 3×10^7
   b) i) So that the agar is not contaminated by other bacteria [1]; in case harmful bacteria are growing on the plate [1]
      ii) To stop condensation falling on the bacteria [1]
   c) i) There are so many colonies [1]; they have joined together [1]
      ii) More [1]; $\frac{3\,000\,000}{1\,000\,000}$ = 3 bacteria [1]; therefore, there should be 3 colonies (but Tilly has 4 colonies) [1]

Page 28 Cell Division
1. a) Cells that are unspecialised / undifferentiated [1]; and can form any type of cell [1]
   b) Stem cells could be used to replace damaged cells [1]; and treat diseases, such as Parkinson's disease / diabetes. [1]; Embryonic stem cells are usually extracted from embryos [1]; which results in the destruction of the embryos [1]; some people have ethical / religious objections to this [1]

Page 28 Transport In and Out of Cells
1. a) particles / molecules [1]; high [1]; low [1]; greater / faster [1]
   b) Active transport moves molecules from low to high concentration [1]; it requires a transfer of energy from respiration [1]
2. a) Concentration of sugar solution [1]

   The independent variable is the one that is deliberately changed.

   b) **Any one of:** all the chips from the same potato [1]; same size chip [1]; length of time in the solution [1]; volume of solution [1]
   c) 0.0mol/dm³ [1]
   d) The potato chip took in water [1]; by osmosis [1]; because the concentration inside the cell was greater than the solution outside [1]
   e) Repeat the experiment for each concentration several times and calculate the mean (average) [1]

3. 1 mark for each correct row **[4]**

	Osmosis	Diffusion	Active Transport
Can cause a substance to enter a cell	✓	✓	✓
Needs energy from respiration	✗	✗	✓
Can move a substance against a concentration gradient	✗	✗	✓
Is responsible for oxygen moving into the red blood cells in the lungs	✗	✓	✗

Page 30 Levels of Organisation
1. Four correctly drawn lines **[3]** (2 marks for two correct lines and 1 mark for one correct line)
   A cell that is hollow and forms tubes – to transport water
   A cell that has a flagellum – to swim
   A cell that is full of protein fibres – to contract
   A cell that has a long projection with branched endings – to carry nerve impulses

Page 30 Digestion
1. a) Liver **[1]**
   b) Gall bladder **[1]**
   c) Small intestine / duodenum **[1]**
   d) To give the best pH for enzymes in the small intestine to work **[1]**
2. a) i) Protease **[1]**
      ii) Amino acids **[1]**
   b) i) The type of enzyme **[1]**
      ii) In each trial, the same concentration of enzyme and mass of protein is used **[1]**

> Try to avoid using the term 'amount'. Use concentration or mass instead.

   iii) 1 mark for each correctly plotted bar **[4]**

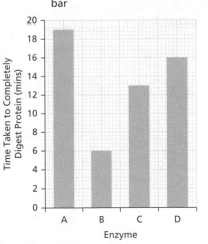

---

c) Enzyme B **[1]**; it digests the protein more quickly that the other enzymes **[1]**
d) Repeat the experiment and calculate the mean (average) of the results **[1]**

Page 32 Blood and the Circulation
1. a) A = trachea **[1]**; B = bronchus **[1]**; C = bronchiole **[1]**
   b) i) Carbon dioxide **[1]**
      ii) Oxygen **[1]**
2. lungs **[1]**; vein **[1]**; body **[1]**; deoxygenated **[1]**; vena cava **[1]**

> Remember: arteries always take blood away from the heart ('A' for away and 'A' for artery).

Page 32 Non-Communicable Diseases
1. Four correctly drawn lines **[3]** (2 marks for two correct lines and 1 mark for one correct line)
   Type 2 diabetes – obesity
   liver damage – excess alcohol intake
   lung cancer – smoking
   skin cancer – UV light
2. a) $89.5 - 38.9 = 50.6$, $\frac{50.6}{89.5} \times 100$ **[1]**; $= 56.5\%$ **[1]**
   b) There are fewer alveoli **[1]**; so less surface area **[1]**; therefore, slower / less diffusion **[1]**
   c) There is less oxygen for respiration **[1]**; therefore, less energy available **[1]**

Page 33 Transport in Plants
1. absorbing carbon dioxide **[1]**; giving off water vapour **[1]**; giving off oxygen **[1]**
2. a) Xylem **[1]**
   b) Phloem **[1]**

Pages 34–55 Revise Questions

Page 35 Quick Test
1. Use of a handkerchief / tissue
2. So that microorganisms from the uncooked meat do not contaminate the cooked meat
3. It is not a barrier method of contraception
4. To prevent the fungal spores spreading to healthy leaves

Page 37 Quick Test
1. With acid
2. The process that a white blood cell uses to engulf pathogens
3. Antibodies
4. To stimulate antibody production without making the person too ill

Page 39 Quick Test
1. Antibiotics only kill bacteria (HIV is caused by a virus)
2. Resistant
3. Because willow bark contains aspirin
4. Neither the patient nor the doctor know which medication the patient is taking

---

Page 41 Quick Test
1. It reduces chlorophyll content so less light is trapped
2. Liquid (containing sugars) from the phloem
3. Nitrates are needed to make proteins

Page 43 Quick Test
1. Chlorophyll
2. From the air
3. **Any three of:** temperature; carbon dioxide concentration; light availability; chlorophyll concentration
4. Nitrate ions (nitrogen)

Page 45 Quick Test
1. Aerobic
2. Respiration rate increases to generate more heat
3. Lactic acid
4. The liver
5. Lactic acid builds up and causes cramp

Page 47 Quick Test
1. Enzymes denature at high temperatures
2. 37°C
3. Increased sweating; vasodilation
4. To generate heat – as muscles contract and relax uncontrollably, heat energy is released

Page 49 Quick Test
1. Sensory
2. To protect the retina from damage
3. A synapse

Page 51 Quick Test
1. In the neck
2. In the pancreas
3. A waste product from the breakdown of proteins containing nitrogen
4. It will reduce the release of ADH

Page 53 Quick Test
1. Testosterone
2. Pituitary gland
3. LH
4. Outside the body / in a Petri dish

Page 55 Quick Test
1. a) Auxins
   b) Causes them to elongate
2. Phototropism
3. They will only kill certain plants (the weeds)
4. The ethene from the red tomato will ripen the green tomatoes

Pages 56–63 Review Questions

Page 56 Cell Structure
1. cytoplasm **[1]**; cell wall **[1]**; free within the cell **[1]**; plasmids **[1]**
2. a) In the ribosomes **[1]**
   b) In the cytoplasm **[1]**
   c) Mitochondria **[1]**

Answers

Page 57 Investigating Cells
1. a) 25mg/dl [1]
   b) As time increased, the concentration of glucose in the cell increased until it reached a maximum [1]; after 120 minutes the concentration in the cell became constant [1]
   c) Glucose is diffusing into the cell [1]; because the concentration is higher outside / down a diffusion gradient [1]; after 120 minutes the concentrations become equal [1]
2. a) 20µm [1]
   b) There are chloroplasts present [1]
   c) **Any one of:** no cell wall [1]; moves around due to flagellum [1]
   d) The chromosomes must divide / DNA must be replicated [1]

Page 58 Cell Division
1. a) $\frac{15}{0.03}$ [1]; = 500 times [1]
   b) Cell A [1]; sets of chromosomes have moved apart [1]
   c) Undifferentiated [1]
   d) i) They are too small [1]; to be seen at this resolution / the resolution of the light microscope is not good enough [1]
      ii) By using an electron microscope [1]

Page 59 Transport In and Out of Cells
1. a) $\frac{-0.6}{2.7} \times 100$ [1]; = −22.2% [1]
   b) No change in mass [1]; which means, no net water uptake or loss / no net osmosis [1]; so, the concentration of the sugar solution is the same as the cell contents [1]
   c) To remove any liquid [1]; so that the mass of the chip is not affected [1]

Page 60 Levels of Organisation
1. Four correctly drawn lines [3] (2 marks for two correct lines and 1 mark for one correct line)
   glandular – can produce enzymes and hormones
   nervous – can carry electrical impulses
   muscular – can contract to bring about movement
   epithelial – a lining / covering tissue

Page 60 Digestion
1. a) i) C [1]
      ii) E [1]
      iii) B [1]
   b) **Any one of:** amylase; lipase; protease; pancreatic juice; sodium hydrogen carbonate; glucagon [1]
2. One mark for each correct enzyme – nutrient – subunit link [3]
   protease – proteins – amino acids
   amylase – starch – maltose
   lipase – fats – glycerol and fatty acids

Page 61 Blood and the Circulation
1. a) Red blood cell / erythrocyte [1]
   b) So that it can fit in more haemoglobin [1]
2. Good blood supply [1]; Large surface area [1]
3. a) i) double [1]
      ii) twice [1]
      iii) arteries [1]
   b) tissues [1]; lungs [1]; glucose / food [1]
4. a) an atrium [1]
   b) an artery [1]
   c) a valve [1]

Page 62 Non-Communicable Diseases
1. a) Build-up of fatty deposits / blockages [1]; in the arteries [1]

Remember to say that that cholesterol build-up happens in the arteries, not veins or just blood vessels.

   b) Less blood reaches heart muscle [1]; less oxygen / glucose available [1]; muscle cells cannot contract / beat [1]; possible heart attack [1]

Page 62 Transport in Plants
1. a) i) 1000 × 10 [1]; = 10 000 [1]
      ii) Y [1]
   b) X [1]; it has fewer stomata [1]; which reduces water loss / transpiration [1]
   c) The lower surface is shaded from the sun [1]; so there is less water loss / transpiration [1]
2. epidermis [1]; mesophyll [1]; gaps [1]
3. transpiration [1]

Pages 64–73 Practice Questions

Page 64 Pathogens and Disease
1. Four correctly drawn lines [3] (2 marks for two correct lines and 1 mark for one correct line)
   bacterium – salmonella
   virus – measles
   protist – malaria
   fungus – rose black spot

Page 64 Human Defences Against Disease
1. dead [1]; weakened [1]; antibodies [1]; immune [1] (first two marks can be given in any order)

Do not use the term 'resistant' here. Resistance is when an organism is born with the ability not to get a disease.

2. a) Fever [1]; red skin rash [1]
   b) It can be fatal [1]
   c) i) 84% [1]
      ii) 1998 [1]; 2003 [1]
      iii) Concern over side effects of the vaccine [1]
3. a) Stops insects biting [1]; so they do not pass on pathogens / diseases [1]; such as malaria [1]
   b) Vaccinations prevent a person from getting a disease [1]; the antibodies need to be in the blood when the pathogen is contracted [1]; otherwise the pathogen may cause illness before antibodies are made [1]
4. Jim [1]; Bill [1]

Page 66 Treating Diseases
1. a) $\frac{(5 + 5 + 6 + 7)}{4}$ [1]; = 5.8 [1]
   b) Plate 4 / bathroom cleaner [1]
   c) To act as a control [1]
   d) She has repeated them four times and calculated the mean [1]
2. a) i) A chemical made by microorganisms [1]; that kills bacteria / stops bacteria reproducing [1]
      ii) Viruses [1]
   b) i) MRSA [1]
      ii) Antibiotics are only available by prescription [1]; only prescribed when necessary [1]; patients are told that they must finish the dose [1]
3. a) 5, 3, 1, 4, 2 [3] (1 mark for one correct; 2 marks for two correct)
   b) To see if they work [1]; to see if they are harmful / have side effects [1]
   c) i) A tablet / liquid that does not contain the drug [1]
      ii) To compare with the effect the drug is having / as a control [1]; to make sure the drug is not just producing a psychological effect [1]

Page 67 Plant Disease
1. a) Phloem [1]
   b) They take food from the phloem [1]; so less is available to the plant for growth / making new cells [1]
   c) The aphids may take up pathogens from the phloem when feeding [1]; and pass them on, via their mouthparts, when they feed on another plant [1]
   d) Tobacco plants contain poisons [1]; which may kill insect pests [1]

Pages 68 Photosynthesis
1. a) water [1]; oxygen [1]
   b) Light [1]; chlorophyll [1]
   c) From the soil [1]
2. a) Use a water bath [1]; make sure the Bunsen is off before pouring the ethanol [1]

Alcohol is flammable, so it should not be directly heated.

   b) The iodine solution will be a yellow / orange colour / no starch is present [1]
   c) To act as a control [1]; to compare with the covered leaf [1]

d) **Any one of:** cover several leaves and test them [1]; test leaves from several plants [1]

Page 68 Respiration and Exercise
1. a) Lactic acid [1]
    b) Aerobic respiration requires oxygen [1]; provides more energy [1]; does not produce lactic acid [1]; produces carbon dioxide and water [1] (or equivalent statements)
2. increases [1]; increase [1]; increases [1]; carbon dioxide [1]; oxygen [1]
3. a) Accurately plotted points [1]; joined by a smooth line [1]

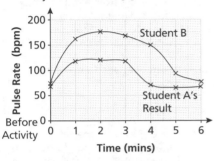

b) Student A [1]; **Any one of:** their pulse rate did not rise so high [1]; their pulse rate returned to normal quicker [1]
    c) Fatigue is caused by a lactic acid build-up [1]; produced by anaerobic respiration [1]; because insufficient oxygen is available [1]

Page 70 Homeostasis and Body Temperature
1. a) i) less [1]
       ii) more [1]
    b) More water will lower the blood concentration [1]; this is detected in the hypothalamus [1]; less ADH will be released [1]; and the kidney will reabsorb less water from the urine [1]

Remember, ADH is 'antidiuretic'. A diuretic makes you produce more urine so ADH makes you produce less.

Page 70 The Nervous System and the Eye
1. a) stimulus, receptor, sensory neurone, relay neurone, motor neurone, effector, response [2] (2 marks for three or four stages in the correct place; 1 mark for two stages in the correct place)
    b) Between the sensory and relay neurone [1]; and between the relay and motor neurone [1]
2. a) Diverging lens ⅠⅠ [1] placed in between the light ray arrow and cornea [1]
    b) Two light rays drawn to show that they spread out after passing through the diverging lens [1]

3. a) Loss of balance [1]
    b) It is difficult to get to the tumours [1]; without damaging surrounding tissues [1]

Page 71 Hormones and Homeostasis
1. Useful substances are returned to the blood [1]
2. a) Liver [1]
    b) Kidney [1]
    c) Amino acids [1]
3. a) selectively permeable [1]; plasma / blood [1]; glucose / amino acids / vitamins [1]
    b) waste products / urea is constantly being made [1]
    c) Make sure the donor organ is a good tissue match [1]; give the patient drugs to suppress their immune system [1]
4. a) 93mg / 100cm³ of blood [1]
    b) 280µg / 100cm³ of blood [1]
    c) 26 minutes after eating the glucose [1]
    d) Increased glucose levels cause release of more insulin [1]; causing greater uptake of glucose into cells [1]; so more glucose is converted to glycogen [1]
    e) The patient's glucose levels would have been higher [1]; and would have stayed higher for longer [1]

Page 72 Hormones and Reproduction
1. a) Ovaries [1]
    b) Possible side effects [1] (Accept named side effects)
    c) Oestrogen inhibits the production of FSH. [1]

Page 73 Plant Hormones
1. a) **Any two of:** use the same type of geraniums [1]; same age geraniums [1]; same soil [1]; same watering [1]; same light [1]
    b) Bob's compound made the geraniums grow quicker [1]; and taller [1]
    c) **Any one of:** it takes longer to prepare [1]; may not always be the same concentration; need access to willow [1]
    d) Auxins / plant hormones [1]

> **Pages 74–93 Revise Questions**

Page 75 Quick Test
1. Long shoots that are produced in asexual reproduction
2. Four
3. 39
4. Sexual reproduction

Page 77 Quick Test
1. DNA
2. A gene
3. Sugar, phosphate and a base
4. A random change in the DNA
5. To switch genes on or off

Page 79 Quick Test
1. Alleles
2. Heterozygous
3. Recessive
4. X and Y

Page 81 Quick Test
1. Darwin (and Wallace)
2. **Any one of:** there was not enough direct evidence; the mechanism for inheritance was not known; it went against the church
3. Lamarck
4. **Any one of:** soft-bodied organisms did not fossilise; fossils have been destroyed

Page 83 Quick Test
1. **Any one of:** high milk yield; fat content of milk; high-quality beef (any other sensible answer)
2. An enzyme
3. Worries about the effects on wild plant populations / human health
4. To stimulate the egg to divide

Page 85 Quick Test
1. Order
2. Three
3. Its genus is *Felis* and species is *(Felis) catus*
4. **Any two of:** changes to the environment over long periods of time; new predators; new diseases; more successful competitors; a single catastrophic event
5. They can no longer mate to produce fertile offspring

Page 87 Quick Test
1. Tropical forest / oak woodland (Accept any other sensible answer)
2. Light; minerals; water; space
3. Organisms that can survive in extreme conditions
4. A square frame used for estimating populations
5. Take more samples and calculate a mean (average)

Page 89 Quick Test
1. Decomposers need water / moisture to decay waste
2. To preserve the crisps, as decomposers need oxygen
3. Through photosynthesis in green plants
4. A feeding level in a food chain
5. The predators run short of food and so some die

Page 91 Quick Test
1. **Any one of:** increasing population; rise in standard of living
2. Acid rain
3. **Any one of:** increased demand for wood; land for crops or farming animals; minerals
4. It increases it because there are fewer trees to take up carbon dioxide by photosynthesis

Answers

Page 93 Quick Test

1. Producing food for current generations while ensuring there are enough resources for future generations
2. **Any two of:** heat; excretion; egestion; uneaten parts
3. Following quotas; only catching larger fish
4. A fungus / *Fusarium*

Pages 94–103 Review Questions

Page 94 Pathogens and Disease

1. a) An organism that carries a disease from one host to another without being infected itself **[1]**
   b) Mosquito **[1]**
   c) Insect repellent / killing the mosquitoes **[1]**; mosquito nets **[1]**

Page 94 Human Defences Against Disease

1. a) dead **[1]**; white **[1]**; antibodies **[1]**; live **[1]**
   b) By becoming infected with the disease **[1]**
2. a) 1250 **[1]**
   b) 8th to 10th May **[1]**
   c) People were not immune **[1]**; they had not been vaccinated / no vaccine was available **[1]**
   d) Viruses are inside cells for much of the time **[1]**; so drugs / antibodies find it more difficult to reach them **[1]**

Page 95 Treating Diseases

1. a) A type of jelly **[1]**; containing chemicals / nutrients that bacteria need to grow **[1]**
   b) To kill any microorganisms **[1]**
   c) To give the bacteria time to grow / be killed **[1]**
   d) i) Disc with the widest clear area labelled, E **[1]**
      ii) Disc with no clear area labelled, P. **[1]**
2. a) A = white blood cell / lymphocyte **[1]**; B = tumour cell **[1]**; C = hybridoma cell **[1]**
   b) Antigen **[1]**
   c) The drug attaches to the antibody **[1]**; the antibody attaches itself to a cancer cell **[1]**; the drug enters the cancer cell **[1]**; and the cancer cell is destroyed **[1]**
3. a) Because the drug may work differently on humans compared with other animals. **[1]**
   b) A placebo is a tablet / liquid containing no drug **[1]**; used for comparison (control) **[1]**; to make sure any positive responses are not just a psychological response to taking a pill / medicine **[1]**
   c) Need to balance the risk with the benefit gained **[1]**; a slight risk of side effects may be acceptable if the benefit is that it relieves great pain **[1]**

Page 97 Plant Disease

1. a) There is less chlorophyll present **[1]**; so less photosynthesis takes place **[1]**; and less food / glucose is made **[1]**
   b) Magnesium ions are needed to make chlorophyll **[1]**
   c) Take infected plant to a laboratory / use a testing kit **[1]**

Page 98 Photosynthesis

1. a) carbon dioxide + water **[1]**; → glucose + oxygen **[1]**
   b) i) Number of bubbles increase as carbon dioxide levels increase **[1]**; then the number of bubbles per minute levels off **[1]**

> Be careful not to say that photosynthesis stops. It is still taking place – it is just not increasing in rate.

   ii) **Any two of:** she kept the lamp the same distance away **[1]**; used the same pondweed **[1]**; kept the temperature the same **[1]**
   iii) There is some carbon dioxide dissolved in the water **[1]**

Page 99 Respiration and Exercise

1. a) i) oxygen **[1]**; carbon dioxide **[1]**
      ii) The muscles need more energy **[1]**; so the respiration rate increases **[1]**; and breathing rate increases to take on board more oxygen **[1]**; lose more carbon dioxide **[1]**
   b) glucose → lactic acid **[1]**
   c) Breathing rate stays high because lactic acid needs to be broken down **[1]**; to pay back the oxygen debt **[1]**; and remove any excess carbon dioxide **[1]**
2. a) Carbon dioxide **[1]**
   b) i) Anaerobic / fermentation **[1]**
      ii) Alcohol / ethanol **[1]**
   c) To provide food **[1]**; for the yeast **[1]**

Page 100 Homeostasis and Body Temperature

1. a) Brain **[1]**
   b) 37°C **[1]**

Page 100 The Nervous System and the Eye

1. a) Sensory neurone **[1]**
   b) A synapse **[1]**
2. a) Diagram showing a smaller pupil **[1]**
   b) i) light **[1]**
      ii) retina **[1]**
      iii) iris **[1]**
3. a) Correct labelled optic nerve **[1]**; cornea **[1]**; and retina **[1]**
   b) Correctly drawn lens **[1]**; and suspensory ligaments **[1]**

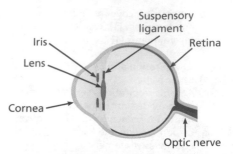

c) The suspensory ligaments slacken **[1]**; the lens becomes more rounded **[1]**
d) The light rays from a distant object are focused in front of the retina **[1]**; the laser is used to change the curvature of the cornea **[1]**

Page 101 Hormones and Homeostasis

1. a) Glycogen **[1]**
   b) Liver **[1]**
   c) Glucagon **[1]**

> Be very careful with the words glucagon, glycogen and glucose. Any spelling mistakes may mean that your answer is unclear and marks may be lost.

   d) Pancreas **[1]**
   e) The pancreas does not produce insulin. **[1]**
2. a) liver **[1]**
   b) amino acids **[1]**
3. a) They are too large to be filtered out of the blood **[1]**
   b) reabsorbed **[1]**
4. a) Accurately plotted points **[1]**; curve of best fit through points **[1]**

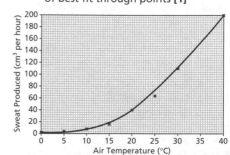

   b) Urine decreases and sweat increases **[1]**; urine decrease is a straight line / sweat production is a curve **[1]**; sweat production is increasing at an increasing rate **[1]**
   c) Sweat increases to try to cool the body down **[1]**; this makes the blood more concentrated **[1]**; so more ADH is released **[1]**; and more water is reabsorbed from the urine (the volume of urine is reduced) **[1]**

Page 103 Hormones and Reproduction

1. a) FSH / follicle stimulating hormone **[1]**
   b) Oestrogen **[1]**

c) i) FSH can be given [1]; to increase production of eggs [1]

ii) Combined pill containing oestrogen and progesterone [1]; inhibits FSH, so ovulation stops [1]

Page 103 Plant Hormones
1. A [1]; D [1]; E [1]; G [1]

Page 104 Sexual and Asexual Reproduction
1. a) A = nucleus [1]; B = cytoplasm [1]; C = chromosomes [1]; D = cell membrane [1]
   b) fusion [1]; DNA [1]; variation [1]
2. a) characteristics [1]; sexual [1]; specialised [1]; implanted [1]; wombs [1]
   b) Many more offspring can be produced [1]; all the offspring will be identical to each other with the desired characteristics [1]
   c) The offspring are identical because they all come from the same zygote [1]; therefore, they have the same genetic information [1]; they have some genetic information from the father [1]; and some from the mother as the original embryo was created using sexual reproduction (sperm and egg) [1]
   d) To obtain stem cells for treating diseases / damaged cells [1]

Page 105 DNA and Protein Synthesis
1. a) Chromosomes [1]
   b) A section of DNA [1]; that codes for a particular sequence of amino acids [1]; to make a specific protein [1]
   c) Double helix [1]
   d) A spontaneous change in the structure of the DNA [1]

Page 105 Patterns of Inheritance
1. a) XX and XY [1]
   b)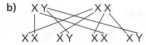
   (1 mark for each correct row [2])
2. Four correctly drawn lines [3] (2 marks for two correct lines and 1 mark for one correct line)
   both alleles are the same – homozygous
   two different alleles – heterozygous
   what the organism looks like – phenotype
   an allele that is always expressed if present – dominant
3. a) NN [1]
   b) Correct parent: nn [1]; correct gametes: N, n, n, n [1]; correct offspring: Nn, nn, Nn, nn [1]
   c) 1 : 1 [1]

Page 106 Variation and Evolution
1. a) The giraffes showed variation [1]; longer-necked individuals were more likely to survive as they could get more food [1]; they breed and pass on the gene / characteristic to their offspring [1]
   b) Changes that happen during an individual's lifetime are not passed on [1]; because they do not change the genetic material [1]
   c) Any two of: There was a lack of direct evidence at the time [1]; the mechanism for inheritance was not known [1]; people were very religious at the time and didn't like that it contradicted the creation story [1]

Page 107 Manipulating Genes
1. a) Enzymes [1]
   b) Any two of: they easily take up genes in plasmids [1]; they use them to make proteins [1]; they replicate rapidly [1]

Page 107 Classification
1. a) Any one of: dead organisms are coved in mud and compressed [1]; from hard parts of animals that do not decay [1]; from parts of organisms that have not decayed because the conditions prevented it [1]; when parts of an organism are replaced by other materials as they decay [1]; preserved traces of organisms, e.g. footprints [1]
   b) May be deep in rock / not many to find as they have been destroyed [1]
   c) Any three of: geographical changes [1]; new predators [1]; new diseases [1]; more competitors [1]; catastrophic events [1]; human actions [1]
   d) Great auk / dodo [1] (Accept any other sensible answer)

Page 108 Ecosystems
1. a) Any two of: extreme temperatures [1]; lack of food [1]; lack of water [1];
   b) i) The thick stem enables it to store water / makes it tough for animals to bite and eat [1]
      ii) Any one of: a smaller surface area so less water loss [1]; stops animals eating it [1]
2. a) Yes, Ben is correct [1]; because the quadrats are not placed at random [1]; they may not be representative of the whole field [1]
   b) Mean = $\frac{(5 + 7 + 9 + 4 + 10)}{5}$ [1]; = 7 [1]; bar with a height of 7 accurately plotted on graph [1]
   c) 9 [1]

d) Horse field [1]
e) 6 × 1000 [1]; = 6000 [1]
f) Grazing reduces the number of plants [1]; the least change is in the number of thistles [1]; the greatest change is in the number of dandelions [1]; horses have a greater impact on the number of dandelions than sheep [1]
g) Increase the number of quadrats [1]
h) Water [1]; minerals [1]; light [1]

Page 109 Cycles and Feeding Relationships
1. a) Top to bottom: herons; frogs; insects; plants [3] (2 marks for two correct; 1 mark for one correct)
   b) Sunlight [1]
   c) Mass is lost between each stage [1]
   d) There would be fewer frogs [1]; as less food would be available to them [1]
2. a) i) A [1]
      ii) D [1]
      iii) B [1]
      iv) C [1]
   b) Combustion [1]

Page 110 Disrupting Ecosystems
1. tropical [1]; trees [1]; carbon dioxide [1]; biodiversity [1]; extinct [1]; habitats [1]

Page 111 Feeding the World
1. a) 100 − 61 − 6 [1]; = 33kJ [1]
   b) 61 + 33 [1]; = 94kJ [1]
   c) Limit the movement of the animals. [1]; Heat the animals' surroundings. [1]

Page 112 Sexual and Asexual Reproduction
1. a) one [1]
   b) gametes [1]
   c) the same as [1]
   d) clone [1]
2. a) Gamete production [1]
   b) chromosome [1]; genetic [1]; parent / old / original [1]; chromosomes [1]
   c) Gametes [1]
3. a) 3, 6, 4, 1 [3] (2 marks for 2 correct; 1 mark for 1 correct)
   b) By giving them an electric shock [1]

Page 113 DNA and Protein Synthesis
1. A gene is a length of DNA [1]; DNA is made up of a sequence of bases [1]; the order of bases codes for the order of amino acids in a protein [1]; three bases for each amino acid [1]
2. a) All the genetic material within an organism [1]
   b) Any two of: doctors can search for genes linked to different types of disorder [1]; it can help scientists to understand the cause of inherited disorders and how to treat them [1] (Accept any other sensible answer)

Answers

Page 113 Patterns of Inheritance
1. a) X   [1]
    b) X or Y   [1]
    c) A boy is produced when a Y sperm from the father fertilises an egg [1]; all the eggs contain X chromosomes [1]

> Make sure you only put one sex chromosome in a sex cell, they can only be X or Y not XX or XY.

2. a)

		White Flower (rr)	
		Genotype of Ovum (r)	Genotype of Ovum (r)
Red Flower (Rr)	Genotype of Pollen (R)	Rr [1]	Rr [1]
	Genotype of Pollen (r)	rr [1]	rr [1]

    b) 12 red [1]; 12 white [1]
    c) This is only a probability [1]; fusion of gametes is random / only gets close to 1 : 1 with large numbers [1]

Page 114 Variation and Evolution
1. Top to bottom: I [1]; I [1]; B [1]; E [1]

Page 115 Manipulating Genes
1. a) A high yield [1]; resistance to disease [1] (Accept any other sensible answer)
    b) A high milk yield [1]; fat content of milk [1] (Accept any other sensible answer)
    c) Choose parents that show the desired characteristic [1]; breed them together [1]; select offspring with the desired characteristic and breed [1]; continue over many generations [1]
    d) Genetically modified / GM   [1]

Page 115 Classification
1. a) i) *Cepaea*   [1]
       ii) Mate them together [1]; see if they produced fertile offspring [1]
    b) Dark-shelled snails are better adapted to cold conditions [1]; so they are more likely to survive and breed [1]; and pass on the dark characteristic to the offspring [1]
    c) Speciation [1]; the populations become isolated so cannot reproduce with each other [1]; natural selection happens differently in the two populations / areas [1]; they become so different they can no longer breed to produce fertile offspring [1]

Pages 116 Ecosystems
1. a) C   [1]
    b) B   [1]
    c) C   [1]
    d) A   [1]
2. a) Place all the quadrats randomly in the field.   [1]
    b) $\frac{(5 + 2 + 1 + 0 + 4 + 5 + 2 + 0 + 6 + 3)}{10}$ [1]; = 2.8 [1]
    c) i) (60 × 90) × 2.3 [1]; = 12 420 [1]
       ii) Area = 3.14 × 5^2 = 78.5m^2 [1]; 78.5 × 2.3 [1]; = 180.55 = 181 (to nearest whole number) [1]

> Area of a circle = πr^2 You are given the diameter in the question, so must divide by two to work out the radius.

Page 117 Cycles and Feeding Relationships
1. algae [1]; decomposers [1]; decomposers [1]

Page 117 Disrupting Ecosystems
1. a) radiation / heat [1]; greenhouse [1]; **Any two of:** carbon dioxide [1]; nitrous oxide [1]; methane [1]
    b) Climate change [1]; A rise in sea level [1]
2. a) 10–15%   [1]
    b) 4% [1]; as it kills many more weeds than 1 or 2 [1]; it is not worth using the extra herbicide for 6% as it has little extra effect [1] (or any other answer supported by a sensible reason)
    c) **Any two of:** the type of weeds present [1]; the type of soil in the fields [1]; amount of watering [1]; amount of light [1]
    d) i) It may get washed into the pond and kill other organisms   [1]
       ii) It may kill the hedges and organisms living in them   [1]

Page 119 Feeding the World
1. a) 1%   [1]
    b) 100 − (30 + 20 + 40) [1]; = 10% [1]
    c) 10% of 1% [1]; = 0.1% [1]
    d) Energy is transferred to the environment when food passes between organisms / 10% of energy is lost at each stage of the food chain [1]; plants are at the bottom of the food chain [1]; so eating vegetables (rather than meat) reduces the number of stages and means less energy is transferred to the environment / more energy is transferred to the person [1]

Pages 120–131 Mixed Exam-Style Questions

1. a) 0.02mm   [1]
    b) Vacuoles [1]; chloroplasts [1]; cell walls [1]
    c) Top to bottom: vacuole [1]; cell membrane [1]; nucleus [1]

2. a) Large surface area [1]; rich blood supply [1]
    b) i) Further increases the surface area [1]; so diffusion is more rapid [1]
       ii) Some substances are taken up by active transport [1]; which requires energy [1]
    c) Less / slower absorption of digested food   [1]
3. a) $\frac{(13 - 6)}{6} \times 100$ [1]; = 116.7% [1]
    b) 90 − 30 = 60 minutes   [1]
    c) It is sent to the liver [1]; where it is broken down using oxygen [1]
    d) To make it a valid experiment / fair test   [1]
    e) The second athlete   [1]
4. a) They are both in the same genus (*Sciurus*)   [1]
    b) Food   [1]
    c) They are not the same species   [1]
    d) Advantage: it should only kill the grey squirrels [1]; Disadvantage: it could kill other animals [1] (Accept any other sensible answers)
    e) Some red squirrels might have the enzymes to digest acorns [1]; they are more likely to compete successfully and survive [1]; they will pass on the gene for making the enzymes [1]; and over generations the population would become able to digest the acorns [1]
5. Concentration of carbon dioxide or the temperature.   [1]
6. The shoot is growing towards the light [1]; showing positive phototropism [1]; the root is growing down in the direction of gravity [1]; showing positive geotropism / gravitropism [1]
7. a) i) Cc   [1]
       ii) cc   [1]
    b) Female sufferer   [1]
    c) Parents: Cc   ×   cc   [1];
       Gametes: C or c   ×   c   [1];
       Offspring: Cc and cc   [1];
       Probability of a baby with cystic fibrosis is 50% [1]
8. a) 210   [1]
    b) Cretaceous   [1]
    c) From their fossils   [1]
    d) Lizard   [1]
    e) Archosaurs   [1]
9. a) Contains a weakened or dead pathogen [1]; stimulates the production of antibodies [1]; by white blood cells [1]; if the normal pathogen infects then antibodies are produced [1]
    b) i) May contain a weakened pathogen [1]; so they get a mild form of the disease [1]
       ii) They might be worried about the side effects [1]; they do not want the children to have three sets of side effects close together [1]

iii) Cheaper / less time-consuming for GPs / worried that they may not come back after first or second vaccination **[1]**

c) To achieve herd immunity / so that the pathogen cannot be passed on **[1]**; to prevent an outbreak **[1]**

10. a) Help the frog to climb / cling onto trees **[1]**
b) Hides it from predators **[1]**
c) To catch insects **[1]**

11. a) Charles Darwin's Theory **[1]**
b) Random change In a gene / DNA / genetic material **[1]**
c) Some of the fossils are of extinct organisms **[1]**; some of the fossils are older than 10 000 years **[1]**

12. a) Accurately plotted points **[2]**; best curve / straight lines drawn connecting points **[1]**

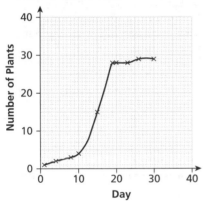

b) Increasing at a steady rate **[1]**
c) **Any two of:** lack of minerals **[1]**; lack of space **[1]**; competition for light **[1]**

d) Take some water from each pond in two separate beakers **[1]**; put the same number of duckweed plants in each sample **[1]**; count the number of duckweed plants in each beaker over time **[1]**

13. a) Four correctly drawn lines **[3]** (2 marks for two correct lines and 1 mark for one correct line)
Increasing the amount of metal recycled – Fewer quarries are dug to provide raw materials
Reducing sulfur dioxide emissions – Less acid rain is produced
Using fewer pesticides and fertilisers – Less pollution of rivers flowing through farmland
Increasing the amount of paper recycled – Fewer forests are cut down

b) Peat decomposes in the gardens **[1]**; and carbon dioxide is given off **[1]** OR Peat bogs are destroyed to obtain the peat **[1]**; which is a threat to plants and animals that live in that habitat **[1]**

c) Producing food without damaging the environment **[1]**; or putting at risk future food supplies **[1]**

14. a) 46 **[1]**
b) Half the number of chromosomes in a human body cell. **[1]**
c) Which sex chromosomes they have **[1]**; XY is male and XX is female **[1]**
d) Parents: XX   ×   XY   **[1]**;

Gametes: X   ×   X or Y   **[1]**;

Offspring: XX or XY   **[1]**

15. a) fungus **[1]**
b) i) Oxygen **[1]**; for respiration **[1]**
ii) Long strands give it the same texture as meat **[1]**
c) i) 2g → 4g → 8g →16g → 32g → 64g (idea of doubling 5 times) **[1]**; 40 hours **[1]**
ii) **Any two of:** produce protein more rapidly **[1]**; suitable for vegetarians **[1]**; contains less fat **[1]**; contains more fibre **[1]**

16. a) Protein **[1]**; glucose **[1]**
b) $\frac{2.00}{0.02}$ **[1]**; = 100 times **[1]**
c) i) Proteins are too large **[1]**
ii) Glucose passes through the filter **[1]**; but is then reabsorbed **[1]**

> You are expected to be able to interpret tables and bar charts showing data about glucose, ions and urea before and after filtration.

17. a) Too cold and chemical reactions would slow down **[1]**; too hot and enzymes would denature **[1]**
b) i) The larger the animal, the less water it has to lose per kilogram of tissue **[1]**
ii) 1cm³ × 500 **[1]**; = 500cm³ **[1]**
iii) It does not have to sweat much **[1]**; so can conserve water **[1]**

Glossary and Index

A

Abiotic   non-living   86

Accommodation   process that occurs to enable the eye to change so that objects at different distances can be focused on the retina   49

Acid rain   rain with a low pH (acidic) due to the gases released by burning fossil fuels   90

Active site   area on an enzyme (lock) that a substrate molecule (key) can fit into   18

Active transport   the movement of substances against a concentration gradient; requires energy   15

Adaptation   the gradual change in a particular organism, over many generations, to become better suited to its environment   87

HT ADH   hormone released from the pituitary gland, which acts on the kidneys causing more water to be reabsorbed back into the blood   51

HT Adrenaline   hormone released from the adrenal gland, which prepares the body for 'fight or flight'   50

Adult stem cells   cells that can differentiate into limited types of cells   13

Aerobic respiration   respiration that uses oxygen to release energy from glucose   44

Agar   a jelly made from algae that is used to culture microorganisms   11

Allele   an alternative form of a particular gene   78

Alveoli   air sacs in the lungs; oxygen diffuses out of them and carbon dioxide diffuses into them   21

Amylase   an enzyme that breaks down starch   19

Anaerobic respiration   respiration that takes place without oxygen, producing energy and lactic acid   44

Antibiotics   medication used to kill bacterial pathogens inside the body   38

Antibody   a protein produced in the body by the immune system to kill specific pathogens   37

Antigen   a marker molecule found on the surface of microorganisms   37

Antitoxin   a chemical released from white blood cells that can neutralise harmful toxins   37

Apex predator   the top consumer in a food chain; has no predators   89

Aphid   a small insect, often called 'greenfly', that sucks food out of the phloem of plants   41

Asexual reproduction   produces new individuals that are identical to their parents; does not involve the fusion of gametes   74–75

Aspirin   a painkilling drug first extracted from the bark of willow trees   39

Atria   the upper chambers of the heart   21

Auxins   a group of growth hormones produced in plants   54–55

B

Benign   a tumour that will not spread to other parts of the body   23

Bile   a fluid, produced by the liver and stored in the gall bladder that emulsifies fat   19

Binary fission   the process that enables a cell to split into two equal-sized, identical cells   11

Binomial system   the method of naming organisms by using their genus and species   84

Biodiversity   the variety of living organisms and the ecosystems in which they live   90–91

Biotechnology   technology that uses living organisms to produce useful products   93

Biotic   living factors   86

Bronchi   the two tubes formed when the trachea divides into two; one passes to each lung   21

Bronchioles   the fine tubes in the lungs that end in alveoli   21

C

Carbohydrase   an enzyme that can break down a carbohydrate   18

Carbon cycle   the constant recycling of carbon through the processes of life, death and decay   88

Catalyst   a substance that increases the rate of a chemical reaction without being changed or used itself   18

Causal mechanism   a factor that makes a disease more likely to occur and the reason for this effect is known   22–23

Cell cycle   the series of growth and division events that a cell goes through during its life   12

Cell membrane   a layer that holds the cytoplasm in a cell and controls what enters and leaves the cell   8

Cell wall   a protective layer, found outside the cell membrane of plant, fungal and bacterial cells, that helps to support the cell   8

Cellulose   a carbohydrate that makes up the cell wall of plants   8

Central nervous system (CNS)   the part of the nervous system made up of the brain and spinal cord   48

Cerebellum   part of the brain, at the rear, that is responsible for balance and controlling movements   49

Cerebral cortex   the area at the top of the brain that is responsible for intelligence and initiating movements   49

Chlorophyll   the green pigment found in most plants; responsible for photosynthesis   42

Chloroplast   a sub-cellular structure containing chlorophyll, which is found in plants and algae that carry out photosynthesis   8

Chlorosis   a condition that means leaves produce insufficient chlorophyll   41

Chromosomes   long molecules found in the nucleus of all cells; made from DNA   12, 76–77

Ciliary muscles   muscles in the eyes that can change the shape of the lens for accommodation   49

Clone   an offspring that is genetically identical to the parent organism   83

HT Collagen   a protein found in connective tissue under the skin and in tendons   77

Communicable   refers to a disease that can be passed on from one person to another   22–23

Competition   occurs when two organisms are trying to obtain the same factors from the environment   86

Concentration gradient   a change in the concentration of a substance from one region to another   14

Contraception   mechanisms that are used to prevent pregnancy occurring as a result of sexual intercourse   53

Cornea   the transparent membrane that covers the eyeball   49

Coronary heart disease   a condition caused by a build-up of fatty deposits in the coronary arteries leading to a lack of blood and oxygen to the heart muscle   23

Culture   a substance that provides the nutrients for the artificial growth of bacteria and other cells   11

Cuttings   a method of asexually reproducing plants, used by gardeners, by planting small lengths of plant shoots   83

Cystic fibrosis   a genetic condition that causes a build-up of mucus in the lungs   79

Cytoplasm   the substance found in living cells (outside the nucleus), where chemical reactions take place   8

D

HT **Deamination** the breakdown of excess amino acids in the liver — 51

Decomposers microorganisms that break down dead organic material — 88

Deficiency disease a disease caused by the lack of an essential element in the diet — 41

Deforestation the destruction of forests by cutting down large areas of trees — 91

Denature when the shape of an enzyme is changed (by excessive temperature or pH) so that it no longer functions — 18

Dialysis the artificial removal of urea and excess material from the blood (used when the kidneys fail) — 51

Diffusion the natural movement of particles from an area of high concentration to an area of low concentration — 14–15

Digitalis a drug extracted from foxgloves, used to treat heart problems — 39

Disease a malfunction / infection of the body — 22–23

DNA nucleic acid molecules that contain genetic information and make up chromosomes — 12, 76–77

Dominant an allele that only needs to be present once in order to be expressed; represented by a capital letter — 78

Double-blind trial a trial where neither the patient nor the doctor know if the patient is receiving the test drug or a placebo — 39

Double circulatory system the type of blood system found in mammals, where the blood goes through the heart twice on each circuit of the body — 21

E

Ecosystem all the organisms that live in a habitat *and* the non-living parts of the habitat — 86

Effector part of the body (e.g. a muscle or a gland) that produces a response to a stimuli — 46

Electron microscope a device that fires electrons at a specimen to obtain a high resolution image — 10–11

Embryonic stem cells cells found in an embryo that can differentiate into any type of cell — 13

Endocrine system a system of glands that release hormones directly into the bloodstream — 50

Endothermic reactions that take in heat / require heat energy — 42

HT **Ethene** a gas that acts as a plant hormone — 55

Eukaryotic cells that have a nucleus and sub-cellular organelles such as mitochondria — 9

Evolution a gradual change in a group of organisms over a long period of time — 80–81

Exothermic reactions that release energy / heat into the environment — 44

Extinct describes a species that has died out — 85

Extremophile an organism that can live in very extreme environments — 87

F

Fermentation the conversion of sugar to alcohol and carbon dioxide in yeast — 44

Fermenter an industrial vessel that is used to grow microorganisms — 93

HT **Fertility drug** a drug that makes it more likely for sexual intercourse to result in pregnancy — 53

Follicle stimulating hormone (FSH) a hormone released by the pituitary gland that causes an egg to develop in the ovaries — 52

Food chain the feeding relationships between organisms — 89

Fossils the remains of animals / plants preserved in rock — 81

G

Gamete a specialised sex cell formed by meiosis — 74

Gene part of a chromosome, made of DNA, which codes for a protein — 12, 76–77

Genetic engineering the process of moving a gene from one organism to another — 82

Genetically modified (GM) organisms that have had specific areas of their genetic material changed using genetic engineering techniques — 82–83

Genome all the genetic material found in an organism or a species — 76–77

Genotype the combination of alleles an individual has for a particular gene, e.g. BB, Bb or bb — 78

Genus a group of closely relate species — 84

HT **Gibberellins** a group of plant hormones — 55

Global warming the increase in the average temperature on Earth due to a rise in the levels of greenhouse gases in the atmosphere — 91

Gravitropic describes a plant's growth response to gravity (also called geotropic) — 54

H

Haemoglobin the red pigment in red blood cells, that carries oxygen to the organs — 20

HT **HCG** a hormone found in urine during early pregnancy — 39

Health the absence of disease *and* a state of complete physical, mental and social well-being — 22

Heterozygous when an individual carries two different alleles for a gene, e.g. Bb — 78

Homeostasis the process of keeping the internal conditions of the body constant — 46

Homozygous when an individual carries two copies of the same allele for a gene, e.g. BB or bb — 78

Hormone a chemical messenger produced by a gland that travels in the blood to its target organ — 50

HT **Hybridoma** a cell produced by the joining of a tumour cell and a white blood cell — 39

Hyperopia the condition that stops a person's eyes from focusing on near objects clearly — 49

I

Immune system the body's defence system against infections and diseases (consists of white blood cells and antibodies) — 36

Immunity the ability to attack a pathogen before it causes disease due to a previous encounter with the pathogen — 37

HT **In vitro fertilisation (IVF)** a process in which an egg is fertilised by sperm outside of the body — 53

Interdependence when one organism relies on another for certain resources / factors — 86

HT **Inverse square law** when a light source is moved to double the distance then the light intensity reduces by a quarter — 43

Iris the coloured part of the eye that changes the size of the pupil in response to different light intensities — 49

L

Lactic acid a compound produced when cells respire without oxygen (i.e. anaerobically) — 45

HT **Limiting factor** a factor that prevents a reaction going any faster — 42–43

Lipase an enzyme that breaks down fat into fatty acids and glycerol — 18–19

Lock and key theory a model used to explain how enzymes work, where the active site is the lock and the substrate is the key   18

Luteinising hormone (LH) a hormone that stimulates the release of an egg in the menstrual cycle   52

M

Magnification how many times larger an image is than the real object   10

Malignant a tumour that can spread to other areas of the body   23

Medulla the area of the brain that controls heartbeat and breathing   49

Meiosis cell division that forms daughter cells with half the number of chromosomes of the parent cell   74

Menstrual cycle the monthly cycle of an egg being released in females; controlled by hormones   52

Meristems areas of cells in plants that can divide to form new cells   13

Metabolism the sum of all the chemical reactions occurring in the body   45

Mimicry when one organism evolves to look or behave like an organism of another species as a defence mechanism   41

Mitochondria the structures in the cytoplasm where energy is produced from chemical reactions   8

Mitosis cell division that forms two daughter cells, each with the same number of chromosomes as the parent cell   12–13

HT **Monoclonal antibody** an antibody produced by a single clone of cells   39

Monohybrid inheritance the pattern of inheritance shown when a characteristic is controlled by a single gene   79

MRSA (Methicillin-Resistant Staphylococcus Aureus) – an antibiotic-resistant bacterium; a 'superbug'   38

HT **Mutation** a spontaneous change in the genetic material of a cell   77

Mycoprotein a protein-rich food produced from fungi   93

Myopia the condition that stops a person's eyes from focusing on distant objects clearly   49

N

Natural selection the survival of individual organisms that are best adapted to their environment   80–81

HT **Negative feedback** a set of events that detects a variable and then correct any change in the variable away from a set value   46

Non-communicable refers to a disease that cannot be passed on from one individual to another   22–23

Non-specific defences the first line of defence against pathogens in general, includes skin, hair, mucus, etc.   36

Nucleotide a molecule made of a phosphate group, a sugar and an organic base   76–77

Nucleus the control centre of the cell   8, 12–13

O

Oestrogen a hormone secreted by the ovaries that inhibits the production of FSH and triggers the production of LH   52

Optic nerve a collection of neurones that pass nerve impulses from the eye to the brain   49

Optimum the conditions at which an enzyme works best   18

Organ a group of tissues gathered together to perform a particular function   17

Organ system a group of organs that all perform related functions   17

Osmosis the movement of water, through a partially permeable membrane, into a solution with a lower water concentration   14–15

Ovulation the release of an egg (ovum) from the ovary into the fallopian tube   52–53

Oxygen debt oxygen deficiency caused by anaerobic respiration during intense / vigorous exercise   45

P

Pacemaker a natural or artificial device that controls heart rate   21

Pathogen a disease-causing microorganism   34–35

Penicillin an antibiotic extracted from the Penicillium fungus   39

Petri dish a round, shallow dish used to grow bacteria   11

Phagocytosis the process by which one cell, such as a white blood cell, surrounds and engulfs another cell   36

Phenotype the physical expression of the genotype, i.e. the characteristic shown   78

Phototropic describes a plant's growth response to light   54–55

Pituitary gland a small gland at the base of the brain that produces hormones; known as the 'master gland'   50

Placebo a dummy drug given to patients during drug trials   39

Plasma the clear fluid part of blood that contains various dissolved substances, such as proteins and mineral ions   20

Plasmid a small circle of bacterial DNA that is independent of the main bacterial chromosome   9

Pollution the contamination of an environment, e.g. by chemicals or water   90

Polydactyly a genetic disorder caused by a dominant allele, where affected people have extra fingers or toes   79

Polymer a large molecule that is made up of many repeating units   76

Population a group of organisms of the same species living together in a habitat   87

Predator an animal that hunts, kills and eats other animals (prey)   89

Prey an organism that is hunted and killed by a predator for food   89

Producer an organism that can make its own food   89

Progesterone a hormone that repairs the lining of the uterus after menstruation and prevents it breaking down   52

Prokaryotic organisms, such as bacteria, that do not have a nucleus or organelles such as mitochondria   9

Protease an enzyme used to break down proteins into amino acids   19

Pulmonary artery the blood vessel carrying deoxygenated blood from the heart to the lungs   21

Pulmonary vein the blood vessel carrying oxygenated blood from the lungs to the heart   21

Punnett square a type of diagram used to work out the outcome of genetic crosses   79

Pupil the opening at the front of the eye that lets light enter   49

Pyramid of biomass a diagram that uses different sized boxes to represent the total biomass at each trophic level in a food web   92

Q

Quadrat a square frame (usually between 0.25m² and 1m²) used for sampling organisms in their natural environment   87

R

Receptors cells found in sense organs, e.g. eyes, ears, nose   46

Refract to bend / change the direction of a light ray (commonly performed by a lens)   49

Recessive an allele that will only be expressed if there are two present; represented by a lower case letter   78

Resolution the smallest distance apart two objects can be and still be seen as separate objects   10–11

Retina the layer at the back of the eye that contains the light-sensitive receptors   49

Ribosomes   small structures found in the cytoplasm of living cells where protein synthesis takes place   8

Risk factor   a factor that will increase the chance of developing a disease   22–23

Runners   long shoots from plants, such as strawberries, that are used for asexual reproduction   74

S

Sclera   the tough white coating to the eye   49

Selective breeding   the breeding process used by scientists and farmers to produce organisms that show the characteristics that are considered useful   82

Selective reabsorption the process of active transport that reabsorbs useful molecules back into the blood from the kidney tubules   51

Sex chromosomes   the pair of chromosomes that determine the sex of organisms   79

Specialised   adapted for a particular purpose   16

Speciation   where populations have become so different that successful interbreeding is no longer possible   85

Species   a group of organisms that can reproduce with each other to produce fertile offspring   84

Statins   a drug used to help lower cholesterol levels in the blood 23

Stem cell   a human embryo cell or adult bone marrow cell that has yet to differentiate   13

Stent   a tube that is inserted into a blood vessel to keep it open 23

Stomata   openings / pores in the leaves of plants   25

Sub-cellular structures   structures found in cells that include the nucleus, mitochondria, chloroplasts and ribosomes   8

Surface area to volume ratio   a way of comparing the surface area of an organism to its volume – the smaller this ratio, the harder it is to exchange substances with the environment at a fast enough rate   14

Surrogate   a female that gestates and gives birth to an organism that has not been produced from one of her own egg cells   83

Suspensory ligaments   structures that attach the ciliary muscle to the lens in the eye   49

Sustainable   resources that can be replaced or maintained in sufficient quantities to support current and future needs   92

Synapse   the gap between two neurones   48

T

Testosterone   a hormone produced by the testes that controls the male sexual characteristics   52

Therapeutic cloning   clones that are produced to treat diseases and will not be allowed to develop into new offspring   13

Thermoregulatory centre   the part of the brain responsible for maintaining a constant body temperature in warm-blooded animals   47

Three-domain system   a new classification system that divides organisms into three domains rather than five kingdoms   84

HT **Thyroxine**   a hormone released from the thyroid gland that controls the metabolic rate of the body   50

Tissue   a group of cells that have a similar structure and function   17

Tissue culture   a method of producing large numbers of plants asexually by growing small part of plants in a nutrient jelly   83

Tobacco-mosaic virus (TMV)   a virus that attacks tobacco and related plants, causing discoloured patches on the leaves   40

Toxin   a poisonous chemical, produced by certain pathogens   35

Trachea   the main tube or windpipe taking air from the mouth down to the lungs   21

Transect line   a fixed line along which sampling of populations, such as species abundance, is measured   87

Translocation   the method by which dissolved food is transported through the phloem in plants   25

Transpiration   the movement of water through a plant from root to leaf   24

Trophic level   a feeding level in a food chain or web   89

Tropism   growth in response to a stimulus, e.g. plants growing towards the light   54–55

Tubules   small tubes in the kidney, where blood is filtered, selective reabsorption takes place, and urine is produced   51

Tumours   groups of cancerous cells   23

Type 1 diabetes   a condition where not enough insulin is produced by the pancreas   51

Type 2 diabetes   a condition where insulin is produced but the cells of the body do not respond to it   51

U

Undifferentiated   a cell that has not yet become specialised   13

Urea   a waste product from the breakdown of proteins formed in the liver and excreted in urine   51

V

Vaccination   a liquid preparation containing inactive or dead pathogens, used to make the body produce antibodies to provide protection against disease   37

Vacuole   a fluid-filled cavity in a cell that is used for storage and support   8

Variation   differences between individuals of the same species   80

Vasoconstriction   occurs when blood vessels in the skin become narrower so that less blood flows close to the surface of the skin 47

Vasodilation   occurs when blood vessels in the skin become wider so that more blood flows close to the surface of the skin to increase heat loss   47

Vector   an organism that carries a pathogen but does not suffer from the disease   34–35

Ventricles   the lower two chambers of the heart   21

W

Water cycle   a series of processes that circulate water through the environment   89

Notes